# THE CHORNOBYL ACCIDENT

*A Comprehensive Risk Assessment*

Victor Poyarkov • Victor Bar'yakhtar

Vladimir Kholosha • Nikolay Shteinberg

Valerie Kukhar' • Vyacheslav Shestopalov

Ivan Los' • Yuri Saenko

EDITED BY

George J. Vargo

## Battelle Press

Columbus • Richland

**Disclaimer**

This report was prepared as an account of work sponsored by an agency of the United States Government. Neither the United States Government nor any agency thereof, nor Battelle Memorial Institute, nor any of their employees, makes any warranty, express or implied, or assumes any legal liability or responsibility for the accuracy, completeness, or usefulness of any information, apparatus, product, or process disclosed, or represents that its use would not infringe privately owned rights. Reference herein to any specific commercial product, process, or service by trade name, trademark, manufacturer, or otherwise does not necessarily constitute or imply its endorsement, recommendation, or favoring by the United States Government or any agency thereof, or Battelle Memorial Institute. The views and opinions of authors expressed herein do not necessarily state or reflect those of the United States Government or any agency thereof.

**PACIFIC NORTHWEST NATIONAL LABORATORY**
operated by
**BATTELLE**
for the
**UNITED STATES DEPARTMENT OF ENERGY**
under Contract DE-AC06-76RLO 1830

Library of Congress Cataloging-in-Publication Data

The Chornobyl accident : a comprehensive risk assessment / authors,
    Victor Poyarkov . . . [et al.] : Edited by George J. Vargo.
       p. cm
    Includes bibliographical references and index.
    ISBN 1-57477-082-9 (hardcover : alk. paper)
    1. Radioactive waste disposal—Risk assessment—Ukraine—
Chornobyl Region. 2. Chernobyl Nuclear Accident, Chornobyl, Ukraine,
1986—Environmental aspects. 3. Chernobyl Nuclear Accident, Chornobyl,
Ukraine, 1986—Health aspects. I. Poyarkov, Victor, 1947– .
II. Vargo, George J., 1956– .
TD898.14.R57C46   2000
363.17′99′0947714—dc21

                                    99-35803
                                      CIP

Battelle Press
505 King Avenue
Columbus, Ohio 43201-2693, USA
614-424-6393 or 1-800-451-3543
Fax: 614-424-3819
E-mail: press@battelle.org
Website: www.battelle.org/bookstore

# CONTENTS

# EDITOR'S FOREWORD

The 26 April 1986 accident at Chornobyl[1] Nuclear Power Plant (NPP) Unit 4 in the Ukraine Republic of the Soviet Union destroyed the reactor core and major portions of the building in which it was housed. In the following steam explosion and subsequent fires, large amounts of radioactive material were released in the form of gases and dust particles. Thermal plumes as high as 10 km (6 mi) dispersed the particles and gases throughout portions of Ukraine, Belarus, and Russia. Significant fallout from this accident was measured in most of Western Europe and traces of contamination from the accident were detected around the world.

Almost 50 years after the communist revolution and in the fifth year of a sharp escalation in Cold War defense spending, this accident was a watershed event. The accident was devastating to the Soviet economy; the direct costs of mitigation and the loss of gross national product wreaked havoc on an already strained economy. In some assessments, the accident is cited as a pivotal factor in the ultimate demise of the Soviet Union [Beschloss and Talbott 1993; Remnick 1993]. It also posed a great challenge to the scientific and technological establishment. In addition, the accident legitimized a fledgling environmental movement which, until that time, had been on the fringe of Soviet society. Abroad, Europe's green movement gained strength and determination in the accident's aftermath. In other industrialized countries, such as Italy, the consequences of the accident caused the leaders to seriously reexamine or terminate their commercial nuclear power programs altogether.

---

[1]In this work, the Ukrainian transliteration "Chornobyl" is used instead of the more familiar Russian transliteration "Chernobyl."

The story of the liquidators (or accident remediation workers), those who stood on the front line of mitigating the early phases of the accident, is both an interesting and unfortunate chapter in Soviet history. Beyond the economic stresses imposed by the Reagan-era Cold War defense spending escalation, the Chornobyl NPP accident was superimposed on a country already in the midst of a protracted and increasingly unpopular war in Afghanistan. While the leadership of the organizations responsible for eliminating the accident's consequences were drawn from the premiere nuclear establishments at Chelyabinsk, Tomsk, and Krasnoyarsk, supported by senior scientists from Kurchatov Institute and other nuclear design and research institutes, the real work was delegated to the historic defenders of the homeland—the Soviet Army. In the timeframe of the accident, young Soviet Army conscripts were faced with one of two grim career paths and opportunities to earn a military decoration: either join the war in Afghanistan or supplement the legion of those assigned to mitigate the consequences of Chornobyl.

There have been a number of accounts written about the Chornobyl NPP accident. Soviet expatriates with political agendas or westerners with little first-hand working experience in the former Soviet Union wrote many of these accounts. Now, a group of leading Ukrainian scientists and specialists who have dealt with the accident and its aftermath on a first-hand basis for the last 13 years offer their unique perspective. These are the natives who remained behind after the breakup of the Soviet Union and the exodus of many scientists and engineers from the nuclear design and research institutes in Russia.

The idea of publishing a native Ukrainian account of the accident, particularly from a risk assessment perspective, originated in the Kiev University office of my friend and colleague Dr. Victor Poyarkov in May 1998. We recognized both the unique perspective of an all-Ukrainian account and risk assessment and the fact that some of the principal players in mitigating the accident and navigating the transition to the post-Soviet era are now approaching retirement.

There is often a fine line between editing the writings of another to improve understanding and assuming the role of a co-author. This was particularly difficult because a literal translation of the original Russian manuscript[2] contained sentence structures that often appeared convoluted and sometimes unintelligible. I faced another dilemma when reviewing the

---

[2]The manuscript was originally prepared in Russian, Ukrainian, and English. While the authors are Ukrainian and the official language is Ukrainian, the working language in nuclear science and technology is nearly always Russian.

radiological analyses proposed by the authors. In some cases, they have taken an à la carte approach, combining the dose limits of ICRP-2 [ICRP 1959], with the later dosimetric concepts of ICRP-26, -30, and -60 [ICRP 1977; ICRP 1979; ICRP 1991]. While a more consistent and rigorous approach is certainly appropriate, the basic conclusions of their analyses for phenomena such as a collapse of the Shelter structure remain unchanged.

In other cases the authors selected citations for risk estimates that may appear to be inconsistent. While I questioned the basis for their selections I left them unchanged.

The authors presented nearly all of their results in SI radiological quantities and units. Historical units are included only in those cases where they appeared in the original manuscript. I have prepared an appendix on radioactivity, radiation, and radiation health effects that may be of interest and benefit to the nonspecialist.

The human health consequences of the Chornobyl accident have been the subject of wild and sometimes absurd claims. During the development of this project we debated the need to include a summary of the medical effects arising from the Chornobyl accident. In order to present what we hope to be a truly comprehensive assessment of the risks from this accident, I have included a brief summary of the documented medical effects as Appendix B. Some of these, particularly the absence of a significant increase in the leukemia rate in the exposed population, call into question some of the current assumptions and bases for human health risk assessment. There are many excellent publications dealing exclusively with the human health risk dimensions of the accident and I have included several of them in the references.

My involvement with the Chornobyl NPP did not begin until nearly the tenth anniversary of the accident in 1996. This was about the same time the original manuscript for this book was nearing completion. Several important developments have occurred since then that deserve mention. In 1996, the U.S. Department of Energy authorized several projects to make significant short-term safety improvements at the Shelter in anticipation of a larger multinational effort to stabilize and transform the Shelter into a structure that met the generally accepted standards for interim safe storage (i.e., SAFESTOR[3]). These projects involved the installation of an improved nuclear safety monitoring system; equipment for dust suppression; industrial personnel safety enhancements; and hardware, software, and training for improving radiation safety. I led the latter of these and spent much of the following 3 years working on developing specifications, procuring equipment,

---

[3]SAFESTOR implies interim safe storage to allow for the decay of radionuclides and is typically on the order of 50 to 75 years.

The editor inspecting the Shelter interior, 1996

and, most importantly, interacting directly with the staff at the Shelter, to build a sustainable radiation safety infrastructure.

In 1997, I encountered an opportunity to lead a project team to evaluate some emergency repairs that were needed to the Unit 3/4 ventilation stack. During the 1986 accident, this structure, common to Units 3 and 4, sustained significant damage. This ventilation stack consists of an external skeletal structure with a flue that is hung from the top of the superstructure. Accident-generated missiles including fragments of the reactor core as large as 50 kg fractured a number of welded joints in the superstructure or caused deep dents in skeletal members that compromised the structural integrity of the superstructure. By 1996, a collapse of the ventilation stack was considered the most significant accident risk at the Chornobyl NPP site.

The Ukrainian construction company responsible for much of the original site construction proposed a repair plan that cost nearly 4 million dollars (US) and 46 person-Sv (4,600 person-rems) to complete. By analyzing the work plans, our team devised a strategy for completing the project while reducing the cost to under 2 million dollars (US) and less than 4.5 person-Sv [Vargo 1997]. To convince our Ukrainian colleagues to adopt our approach, we had to challenge a number of social factors, such as the expectation of using a large disposable work force, and historical site practices, such as maintaining strict separation between operations at Units 3 and 4. In the end, the work planning and radiation protection review that we performed was cited by senior management at the Shelter as a benchmark for future projects.

In 1997, the G-7[4] Nuclear Safety Working Group recommended a phased systematic approach to long-term management of the Shelter. Unlike some previous proposals for long-term remediation of the Shelter, this effort included substantial Ukrainian participation. This approach, subsequently

---

[4]The G-7 countries are the United States, Canada, the United Kingdom, Germany, France, Italy, and Japan

called the Shelter Implementation Program (SIP) [Heriot and Belusov 1997] addresses five major goals:

- Reduce the potential for collapse of the Shelter
- Reduce the consequences of a Shelter collapse if one should ever occur
- Improve the nuclear safety of the Shelter
- Improve worker safety and environmental protection
- Develop a long-term strategy and study for conversion into an environmentally safe site.

The SIP is not an explicit technical solution but rather a systematic approach to finding an optimized solution. It consists of 22 primary tasks and 257 constituent activities. Some of these projects and tasks were required before any large-scale remediation activities. These projects and tasks were grouped into four so-called early biddable projects. Also, the SIP provides a basis for projecting schedule and cost. The best current estimates are 8 to 9 years until completion at a cost of approximately $750 million dollars (US). Despite this, there still remains considerable uncertainty in the estimates for several reasons:

- Presence of radiological and physical hazards
- Lack of reliable data, particularly on the structural integrity of components and contents of the Shelter
- Need to take an exploratory approach to projects rather than a predefined technical approach
- Lack of previous experience and the unique nature of the Shelter
- Challenge of integrating existing and prior projects with the Shelter Implementation Program.

Execution of the SIP will be accomplished by an international project management unit composed of specialists from Western construction and technology firms that are integrated with staff from the Chornobyl Shelter organization. The SIP does not address the fate of the lava-like fuel-containing masses. Given the lack of any high-level waste management strategy or facilities in Ukraine, the most cost-effective and rational approach is to ensure safe interim storage of these materials in place and to make adequate preparations for their eventual removal and disposition.

The editor inspecting lead blankets destined for remediation work at the Shelter.

The SIP also provides a basis for regulatory review and licensing activities. Because of its unique nature, existing regulations and standards for nuclear safety and waste management are not relevant or applicable to the Shelter. The State Nuclear Regulatory Authority issued a license for the Shelter in 1997. This license requires that the Shelter organization takes actions necessary to prevent degradation of the Shelter, prevent or mitigate uncontrolled releases of radioactive material, and obtain regulatory review and approval for nonroutine activities. The license requires approval of work plans, risk assessment and mitigation plans, as well as collective radiation dose estimates.

In 1997, Ukraine revised its standards for protection of workers against radiation, adopting a variant of the system of dose limitation based on ICRP-60 [ICRP 1991]. Instead of the standard of 100 mSv (10 rems) more than 5 years, workers will be limited to 20 mSv (2 rems) per year, half of the existing 40 mSv (4 rems) per year limit currently observed in Ukraine. Implementation of these new regulations will present even greater challenges to achieve the objectives of the Shelter Implementation Program.

GEORGE J. VARGO, PH.D., CHP
Richland, Washington
June 1999

## REFERENCES

Beschloss, M.R. and Talbott, S. 1993. *At the highest levels.* Little, Brown and Company, Boston.

Heriot, I.D., and E.L. Belusov. 1997. *Shelter implementation plan: a way forward for Chernobyl 4.* 42/519 (10-12), Nuclear Engineering International (October).

International Commission on Radiological Protection (ICRP). 1959. *Report of Committee II on permissible dose for internal radiation.* ICRP Publication 2, Pergamon Press, Oxford.

International Commission on Radiological Protection (ICRP). 1977. *Recommendations of the International Commission on Radiological Protection. Annals of the ICRP 21(1-3).* ICRP Publication 26, Pergamon Press, Oxford.

International Commission on Radiological Protection (ICRP). 1979. *Limits for intakes of radionuclides by workers. Annals of the ICRP 2(3-4).* ICRP Publication 30. Pergamon Press, Oxford.

International Commission on Radiological Protection (ICRP). 1991. *1990 recommendations of the International Commission on Radiological Protection. Annals of the ICRP 1(3).* ICRP Publication 60, Pergamon Press, Oxford.

Remnick, D. 1993. *Lenin's tomb.* Random House, New York.

Vargo, G.J. 1997. *Foreign trip report: travel to Ukraine to evaluate the collective dose estimate and schedule for repairs to the Chornobyl Unit 3/4 ventilation stack, Kyiv Region 25.* Pacific Northwest National Laboratory, Richland, Washington (May and June).

# ACKNOWLEDGEMENTS

First, I want to thank Laurin Dodd, Dennis Kreid, and Bryan Gore at Pacific Northwest National Laboratory and Terry Lash at the US Department of Energy's Office of International Nuclear Safety and Cooperation for the opportunity to work on improving worker safety at the Chornobyl Nuclear Power Plant. It was through this project that I met the authors of this book and established many personal and professional relationships with colleagues and counterparts in Ukraine.

Any project of this type requires the help and support of a variety of people. Thanks to Kevin Hendzel, Igor Kadenko, Paul Makinen, Renata Dunn Morriso, Nikolay Petrovsky, and Vyacheslav Varetsky for translating the original Russian manuscript. The large number of co-authors, different writing styles, and specialized terminology complicated their task.

Kristin Manke, Senior Communications Specialist at Battelle, provided invaluable assistance with transforming the sometimes fractured English into crisp understandable prose. Terri Gilbride also added considerable value through her questions, comments, and suggestions.

Special thanks go to Ron Kathren and others who reviewed the final draft.

Finally, thanks to Joe Sheldrick at Battelle Press for his enthusiasm, encouragement, and patience with making this project a reality.

# I
## ВВЕДЕНИЕ

# INTRODUCTION

V. Poyarkov

*Risk: a chance of encountering harm, loss, danger, etc.*
Webster's Dictionary, 4th edition

Looking at other definitions of risk, we realize just how subjective the concept of risk actually is [Kaplan and Garrick 1987]. The concept of risk is different for each person, institution, and organization; while a certain potential outcome might be viewed by one person as a significant risk, another might view it as insignificant. For example, some would view the loss of $10,000 as total ruin, while others might consider it an insignificant loss. To some extent, all people take risk into account when making decisions on a daily basis. These decisions assume the existence of voluntary and involuntary risk. Risk assessment is important for several reasons, but there are two especially important reasons. First, risk assessment is used for identification of high-priority tasks in the risk management process and for making key health policy decisions. Second, the public should be informed of the risk factors related to environmental conditions. Risk assessment involves a quantitative or qualitative description of the deleterious impact on the health of individuals or population groups resulting from the human effects of various types of environmental damage. In this book, we will discuss the risks brought to the world as a result of the accident at the Chornobyl Nuclear Power Plant (NPP) in Ukraine.

The accident at Chornobyl NPP on 26 April 1986 destroyed reactor Unit 4 and released approximately $11 \times 10^{18}$ Bq (or 1,100 PBq) into the environment (Figure 1.1). Approximately $600 \times 10^{15}$ Bq of long-lived radionuclides remained in the destroyed reactor. Moreover, approximately $8 \times 10^{15}$ Bq of high-level, medium-level, and low-level radioactive waste from the

**FIGURE 1.1.** On 26 April 1986, the accident at Chornobyl Nuclear Power Plant completely destroyed reactor Unit 4 and released approximately $11 \times 10^{18}$ Bq into the environment.

Chornobyl NPP accident is currently stored at more than 800 interim storage sites. Approximately 30,000 $km^2$ of land in Belarus, Ukraine, and Russia is now contaminated in excess of 185 $kBq/m^2$; because of this high contamination, many of the residents on this land were forced to leave.

This radioactive contamination imposes an additional risk on the people in the regions affected. How great is this risk? What is the potential loss of life and health hazard posed by a collapse of the Shelter erected over the destroyed reactor? By radionuclide migration from burial facilities? By radionuclide transfer from soil to vegetation? What is the likelihood of these processes?

Answers to these questions will serve as a scientific foundation for development of countermeasures against the Chornobyl NPP accident and nuclear accidents in general. Such information is also important in explaining the actual hazards associated with nuclear accidents. We believe that the information presented here will act as a counterweight to the public view of the risk created by the initial suppression of the accident and several subsequent publications that unreasonably overstated the potential negative impact.

This report presents a comprehensive analysis of the risks associated with the Chornobyl NPP accident. The analysis begins with a reconstruction of the events that led to the explosion and a discussion of causes. The analysis shows that the most likely cause for the reactor core explosion was the insertion of positive reactivity produced by displacement units on the ends of the reactor control and protection system control rods (see Chapter 2).

On 26 April 1986, Unit 4 experienced an uncontrolled power surge from an increase in reactivity caused by the positive reactivity provided by the displacement units on the reactor control and protection system control rods. Because of the reactor's large positive void coefficient—meaning the nuclear chain reaction and power output increased when cooling water was lost or displaced from the reactor's channels—this uncontrolled increase in power was followed by a rise in power distribution nonuniformity and an explosion in the lower portion of the reactor core. The explosion was equivalent in energy to between 30 and 40 metric tons of trinitrotoluene (TNT).[1] This explosion and subsequent fire destroyed Unit 4 and dispersed $11 \times 10^{18}$ Bq of radionuclides into the environment.

> **MAGNITUDE OF THE ACCIDENT**
>
> ◆ Three people died during the accident on 26 April. Twenty-eight more people died from the effects of radiation in the weeks and months after the accident.
>
> ◆ Approximately 116,000 people were temporarily (and in many cases permanently) relocated from their homes because of the radioactive contamination.
>
> ◆ The accident released $11 \times 10^{18}$ Bq of radiation into the environment.
>
> ◆ Approximately 20–50% of the radioactive iodine in the reactor core at Chornobyl NPP Unit 4 was released. At the Three-Mile Island accident in the United States, less than $2 \times 10^{-5}$% was released.
>
> ◆ The radiation released from the Chornobyl NPP accident was detected at monitoring stations around the world.
>
> ◆ To reduce the radiation released into the environment, a Shelter was built around the destroyed reactor. Approximately 117,000 workers built the Shelter; the collective dose was 9,800 person-Sv.

In the months following the accident, the Shelter was constructed over the destroyed reactor to minimize the spread of airborne contamination. The Shelter was built quickly, under extreme conditions, to provide immediate, temporary containment. Now, 13 years later, there are significant concerns about the structural integrity and life expectancy of the Shelter.

---

[1]Editor's Note: The energy released from trinitrotoluene is a standard measure of explosive force, with a release of 1 kg of TNT equal to approximately 4,600 kJ.

An analysis of the integrity of the Shelter was performed along with an evaluation of potential scenarios for structural failure or collapse. These permit an assessment of the risk associated with a potential release of radioactive dust from the destroyed reactor, as well as subsequent atmospheric transport of such dust.

Several of the physical and chemical forms of the radioactive fallout generated in the 1986 accident were also studied, enabling us to determine the behavior of the fuel particles as a function of time in various types of soil and to predict the soil-to-vegetation transfer characteristics of the radionuclides.

Reconstruction of the radiation environment during and after the accident enabled us to estimate the internal and external doses received by various categories of the public and accident cleanup personnel. This in turn made it possible to compare the risk posed by accident-related radionuclides to that of natural, "non-Chornobyl" radiation (radon, potassium, etc.)

In addition, an inventory of radioactive waste disposal sites and interim radioactive waste storage sites inside the 30-km Exclusion Zone (the area near the Chornobyl NPP evacuated in 1986) was prepared. This enabled us to obtain the input data required to estimate the risk associated with radioactive waste in the Exclusion Zone.

However, scientifically based risk assessments do not always correlate with the public's perception of risk. The authors believe that the results of their studies, collected here, can be used to optimize a strategy for minimizing the negative impact of the Chornobyl NPP accident and to improve the response to nuclear and radiological accidents.

# 2
## аварии

# THE ACCIDENT
## *Chronology, Causes, and Releases*

V. Bar'yakhtar, V. Poyarkov, V. Kholosha, and N. Shteinberg

The Chornobyl Nuclear Power Plant (NPP) is located next to the Pripyat River in northern Ukraine, approximately 130 km north of Kiev. The site is near the Russian and Belarussian borders. Before the accident there were four operating RBMK-1000 reactors at the site (Units 1 through 4). The first two units were built between 1970 and 1977. The last two units were completed in 1983. Two more reactors were under construction at the site when the accident in Unit 4 occurred on 26 April 1986 [NEA 1995].

## 2.1 RBMK-1000 ENGINEERING SPECIFICATIONS AND THE ACCIDENT

Unit 4 contained a RBMK-1000 boiling water reactor and was capable of generating 1000 megawatts of electricity ($MW_e$). The reactor fuel was uranium dioxide enriched to 2–3% $^{235}U$. In the reactor, water is pumped to the bottom of the channels where the uranium fuel (clad in zirconium alloy) is located. The water boils as it progresses up the pressure tubes, producing steam which feeds two turbines. Raising or lowering 211 control rods manages the reactivity or power of the reactor. When lowered, the rods absorb neutrons and reduce the fission rate, thus, reducing the reactor's power. The absolute minimum requirement for safe operation of the reactor is 30 control rods fully inserted (or the equivalent) [NEA 1995].

No industrial facility has ideal specifications. The development and operation of an industrial facility such as a NPP always involves a cost-benefit tradeoff. Unfortunately, the struggle to reap maximum benefit at minimum cost frequently fails to provide benefits over the long term. The omission of important safety studies and safety measures for the RBMK NPP design had tragic consequences.

The design for Chornobyl NPP Units 3 and 4 (Phase II) was developed by the Project Design Engineer in 1974 and approved in consultation with the Scientific Project Manager and Lead Design Engineer. Included in the design were several substantial deviations from safety requirements in force at the time. The effect of these deviations on safety was not analyzed. The RBMK design clearly had at least two weak points:

- A positive void coefficient, meaning the nuclear chain reaction and power output increased when cooling water was lost or displaced from the reactor's channels[1]

- A reactor protection system that was too slow in shutting down the reactor for certain operating modes and that could, under certain circumstances, actually cause a transient power spike (sometimes called a "positive scram").

These weak points can be traced back to the deviations from safety regulations and standards, as well as the incomplete safety studies.

The RBMK reactor control and protection system (RCPS) was designed to compensate for reactivity effects such as

- Loss of fuel channel coolant water in a cold reactor

- Reactor-core steam collapse when fuel elements are cooled to 265°C

- Possibility that one or more reactor protection control rods might stick.

This set of reactivity effects does not cover many of the potential problems that were recognized even during the early phases of reactor operation. For example, the fact that the power coefficient and void coefficient each vary over a wide range of negative and positive values, depending on the composition of the reactor core and the reactor operating mode, was not considered. Further, the fact that the RCPS control rod design forces the introduction of positive reactivity when the control rods are initially inserted into the reactor core from the topmost position (the "positive scram") was not considered. This positive reactivity is due to the fact that the graphite displacement unit on the leading end of the control rod provides less moderation than the water displaced from the lower part of the channel. Thus, a

---

[1]Editor's Note: In general, graphite reactors have more neutron moderator than necessary. This condition is called "overmoderation." In an overmoderated system, the excess moderator sometimes functions as a neutron absorber, reducing the number of neutrons available to sustain the fission cycle. In an overmoderated RBMK reactor, a loss of coolant will cause the power produced by the reactor to increase.

positive reactivity is introduced into the lower portion of the reactor core when the RCPS control rod is moved downward from the upper limit. All RCPS control rods, including the reactor protection control rods, required the same amount of time for insertion into the reactor core (18–21 s). This speed was inadequate for a reactor having large positive feedback. The slow response of the reactor protection system, coupled with the design flaw in the rods (leading to an initially positive increase in reactivity) meant that in some reactor modes, the reactor protection system could initiate a reactor excursion itself.

The RBMK-1000 reactor vault was not protected against overpressurization resulting from multiple fuel-channel leaks. Thus, if several (i.e., approximately 6 of the 1,600) fuel channels developed a leak, the upper reactor plate (System E, No. 8 in Figure 3.1.2, also referred to by some as the "Elena") could be blown off, thereby removing the RCPS control rods from the core and adding positive reactivity. This effect substantially increased the impact of the 26 April 1986 accident.

Almost all of the design flaws in the RCPS were known before the accident. The RBMK-1000 RCPS design flaws could be traced to violations of the following safety-related regulations: Sections 3.1.6, 3.1.8. 3.2.2, 3.3.1, 3.3.5, 3.3.21, 3.3.26, 3.3.28, and 3.3.29 of *Nuclear Safety Regulations for Nuclear Power Plants* (PBYa-04-74) [Atimizdat 1976] and similar requirements contained in Sections 2.2.5, 2.2.6, 2.2.7, 2.2.8, 2.5.2, and 2.5.8 of *General Safety Provisions for Nuclear Power Plants During Design, Construction, and Operation* (OPB-73) [Atomizdat 1974]. The measures required to eliminate the design flaws were also well known, but only began to be implemented after the accident.

Reliable power control (control over the rate of the chain reaction) is extremely important in unstable facilities, such as the pre-accident RBMK reactors. Power control was provided by two systems—the power distribution monitoring system and the RCPS. The power distribution monitoring system has sensors located within the reactor core; the RCPS has sensors located both in the lateral biological shielding tanks and in the reactor core. These two systems are complementary, but each has substantial deficiencies, primarily at low power. The deficiencies are due to the fact that the power distribution monitoring system supports monitoring of the relative and absolute power distribution over 10–120% of nominal power and monitoring of total reactor power over 5–120% of nominal power. The local automatic control and protection system, which responded to signals from the side ionization chambers, was used for reactor control at power below 10% $N_{nom}$, the design-basis power output of the reactor. In the case of the RBMK-1000 reactor, $N_{nom} = 1,000$ MW$_e$.

It is quite difficult to control a reactor the size of the RBMK-1000 (with a core 11.8 m in diameter and 7.0 m high) at low power using side-mounted ionization chambers. At low power, side-mounted ionization chambers cannot "see" into the center of the reactor core and also cannot "see" the vertical power distribution in the reactor core, because the ionization chambers are all located at the same height around the reactor core.

At low power, the operator is "blind" to local power distribution and possible reactivity anomalies; the operator works more on the basis of experience and intuition than instrument readings. Operating the RBMK-1000 in "blind" mode is marginally acceptable when starting up an unpoisoned reactor, where the power distribution field can be controlled on the basis of a previously performed calculation. However, when shutting down a nonuniformly poisoned reactor, the same low-power operating mode may create large field fluctuations and cause critical nonuniformities in the axial (i.e., vertical) and radial power distribution in the reactor core. This effect was not taken into account before the accident, and unfortunately, no restrictions were imposed with respect to low-power operating conditions.

The failure to provide the operator with sufficient objective information or automatic protection systems for various safety-related parameters such as water flow rate in fuel channels, shutdown margin, etc., is a severe design deficiency. This deficiency forces the human operator to perform the functions of an automatic protection system with respect to various critical safety-related parameters.

Excessive reliance was placed on human performance and administrative controls rather than engineered safety features and automatic control systems.

Another point for consideration is that Chornobyl NPP Unit 4 did not have a reactor containment system. Under the Chornobyl NPP Phase II design, a portion of the primary coolant loop was located outside of the hermetically sealed confinement zone surrounding the reactor and associated major equipment. The rooms containing this portion of the loop were provided with special blowout panels to support controlled discharge of radioactive steam and gas into the atmosphere in the event of a pipe break involving a diameter up to 300 mm. Such an accident was projected to result in a 2.1-rem infant-thyroid dose at the boundary of the restricted sanitary zone[2] (because of inhalation of radioactive isotopes of iodine). More severe accidents, including those involving significant fuel damage, were

---

[2]Editor's Note: Under Soviet safety regulations, a sanitary restricted zone of 3 km radius was established around each NPP. This area nominally corresponds to the Exclusion Area Boundary concept used in siting NPPs in the United States.

considered to be "beyond design basis," and no attempts to address them or mitigate their consequences were made.

Discussing the potential for significant reduction in the effects of severe accidents similar to the 26 April 1986 accident with a containment vessel in place is of little value as no serious research has been conducted in this area. However, the lack of a full-fledged accident containment system in the RBMK-1000 reactors does indicate—and this is important to note—that the "defense in depth" safety philosophy for NPPs was not popular in the Soviet Union. The defense in depth philosophy encourages the use of independent multiple fission product barriers and redundant safety systems for reactivity control and adequate core cooling. Tragically, the Chornobyl NPP disaster reaffirmed the validity and necessity of the defense in depth principle.

## 2.2 EVENTS PRECEDING THE ACCIDENT

Over the history of commercial nuclear power, two accidents involving substantial core damage have occurred at NPPs: one at Three-Mile Island NPP in the United States on 28 March 1979 and one at Chornobyl NPP in the former Soviet Union on 26 April 1986. These accidents did not have a comparable impact. Although the Three-Mile Island accident did lead to a precautionary evacuation, the accident itself did not have any substantial impact on the environment or personal health; however it did result in considerable psychological stress and public outrage. This accident prompted both operating organizations and Western regulatory authorities to perform additional safety studies of nuclear facilities, identify any deficiencies, and eliminate them as soon as possible. Unfortunately, in the Union of Soviet Socialist Republics (USSR), this accident was virtually ignored. In 1984, Academician A. Aleksandrov, President of the USSR Academy of Sciences and Director of the Kurchatov Institute, stated in *Pravda* that "this accident could only have happened in a capitalist country, where profit is more important than safety."

No other accident approaches the dimensions of the 26 April 1986 disaster at Chornobyl NPP in terms of radiological release, acute radiation health effects, and socioeconomic impact. It had an enormous psychological impact on the population of the entire world. The Chornobyl accident and its impact became the subject of wide-ranging politically and propagandistically driven speculation. This speculation exacerbated the impact of the accident, especially in the countries of the former Soviet Union, which suffered the most severe effects of the accident.

The accident occurred during a test that was designed to see how the plant would respond to a turbine/generator trip concurrent with a loss of offsite power. The objective of this test was to determine whether residual energy (i.e., coastdown) in the turbines would be sufficient to supply electricity to essential plant equipment and maintain adequate coolant flow through the reactor core until the diesel emergency power supply came online. This was a generic safety issue identified by the Research and Design Institute for Power Engineering in Moscow—the chief design organization for the RBMK reactor. This feature—the ability to sustain the reactor coolant pumps using residual energy—was included in the second-generation RBMK reactors (Chornobyl Unit 3 and 4, all of the units at Smolensk NPP, and Kursk NPP Units 3 and 4) but had not been tested at Chornobyl Unit 4. The proposed tests were essentially treated as electrical equipment tests. Insufficient analysis was performed to determine the impact of the tests on the reactor.

By current standards, it was inaccurate to classify these tests as purely electrical in nature, as they were expected to cause significant changes in the thermal and physical characteristics of the reactor and required intervention on standard safety systems and interlocks. These tests should have been classified as unit-level integrated tests. The test plans should have been approved in consultation with the Project Design Engineer, the Chief Design Engineer, the Scientific Project Manager for RBMK NPPs (Kurchatov Institute, Moscow), and governmental oversight authorities. However, the regulations in force at the time of the accident [Atimizdat 1976; Atomizdat 1984] did not require nuclear plant management to obtain such approval in consultation with these organizations.

## 2.3  **THE ACCIDENT**

To better understand the accident chronology, a brief description of the characteristics of Chornobyl NPP Unit 4's operating regime is provided. The axial power distribution in the reactor core had two peaks, with the highest neutron fluxes occurring in the top portion of the reactor core (as expected for a reactor with a spent core, control rods in the upper position, and greater xenon buildup at the center than at the edge of the reactor). This distribution has extremely low kinetic stability when combined with the RCPS design then in use. Under the thermal and hydraulic conditions in the reactor core, the coolant remained well below the boiling point and therefore had a low steam content. The steam was only present in the top of the reactor core (under the conditions existing at the time, a small increase in power could lead to a much larger increase in steam content in the bottom of the reactor core than in the top). The thermal and hydraulic parameters of the reactor core before the start of the test caused the reactor to be much more susceptible to self-induced nuclear excursion processes in the bottom of the reactor core.

### 2.3.1 **Chronology of the Accident**

The chain of events leading up to the accident began at 01:06[3] on 25 April 1986 when Unit 4 was placed into maintenance status and preparations for plant testing and shutdown began. While no events occurred until the following day, several procedural deviations occurred that epitomized operators' and managers' attitudes towards safety.

The major events started unfolding at 00:28 on 26 April 1986 when it was ordered that reactor power be reduced to 700 MW$_t$, the power level required for the test. This was achieved at approximately 01:00 on 26 April 1986. To perform the test, the reactor power needed to be reduced to 700 MW$_t$. As a result of the in-growth of $^{135}$Xe, a major source of negative reactivity, the reactor power "collapsed" to approximately 30 MW$_t$.

To compensate for the negative reactivity, several inappropriate actions and efforts were taken. These included removing several of the rods controlling the operating reactivity margin. Removing the control rods increased the reactivity and caused the power to increase to about 200 MW. The reactor was now in a state where the protection system could no longer guarantee termination of the nuclear reaction. The critical event, as viewed through the lens of subsequent events, was the decision to increase reactor power again following the collapse to approximately 30 MW$_t$.

Between the initial power increase following the collapse and stabilization of power output from the unit at 200 MW (which occurred at approximately 01:23), normal procedures were followed, and preparations for the test continued. Immediately before the test (at 01:23), Unit 4 was described as follows:

- Power output 200 MWt

- Operating reactivity margin (value as of 01:22:30 recovered by a post-accident calculation using the PRIZMA-ANALOG software package)

- 8 manual control rods inserted (standard operating orders call for a minimum of 30 control rods)

- Power distribution field with two peaks (with the top peak being larger)

- Coolant flow rate 56,000 m$^3$/h

- Feedwater flow 200 metric tons/h

- Thermal and physical parameters nearly stable.

At 01:22:30, the SKALA central monitoring system produced the last recording of unit parameters on magnetic tape before the accident.

---

[3]Editor's Note: All times are reported in 24-h (military) format.–

The command to close the turbine main steam stop valves was issued at 01:23:04. Over the approximately 30 s required for the four main reactor coolant pumps to run down, the unit remained under control, within established limits for the operating mode in question, and no personnel action was required. At this point, neither the reactor power nor any other reactor parameter required manual or automatic intervention.

At 01:23:40, the operator pushed the AZ-5 button to lower the control rods to shut down the reactor. The reason for this action was never clearly established.

Lowering the control rods substantially deformed the power distribution. In the top of the reactor core, where the control rods were inserted, the neutron flux began to drop. In the lower portion of the reactor core, where the water columns that had served as neutron moderators were being displaced by the graphite spacers on the ends of the control rods, the neutron flux began to rise.

The total reactor power was observed to increase by a factor of several dozen over the initial value within approximately 5 s. The power density distribution form factor, which characterizes homogeneity (or the lack thereof) of energy detachment in the reactor core, reached a value of $K_v = 5.5$. When the initial reactor power (based on the same PRIZMA calculations) increased by approximately a factor of 30, the linear thermal loads in the highest stress areas were much higher than the nominal values at 100% power.

The vertical peak of the power distribution moved 2 m closer to the bottom of the reactor core. The linear thermal loads on the lower portion of the fuel more than doubled, with different areas around the reactor core cross-section being affected differently. In the bottom portion of some fuel channels, the fuel enthalpy reached the critical values for fuel dispersal.[4] The fuel dispersal led to an increase in the surface area for heat exchange between hot fuel particles and the heat transfer agent in the bottom of the channels.

As water came into contact with the dispersed fuel, steam rapidly began to form, increasing pressure in the corresponding fuel channel sections, and, finally, damaging the fuel channel.

Simultaneously, the power density increase caused by inserting the control rods triggered a positive void coefficient, which is an intrinsic property of the RBMK-1000.

When the thermal inertia of the fuel elements was overcome, steam began to form in the areas of highest power density. The onset of and increase in

---

[4]Editor's Note: Although the authors cite 285 cal/g as the threshold for rapid dispersal of uranium dioxide, experimentally derived estimates vary. U.S. safety analyses use 280 cal/g. The threshold for fuel dispersal is 320 cal/g.

steam formation caused the increased power density to spread to the entire reactor core.

Once 3–4 of the 1,600 fuel channels were damaged, the pressure in the reactor vault increased, literally to the point that the pressure began moving internal equipment. The 1,000-metric-ton biological shield (System E, Figure 3.1.2) was blown off, causing all of the control rods to jam at the halfway point. The frozen readings on the selsyn meters used to indicate control rod position could be seen on the Unit 4 reactor control panel for several years following the accident.

Once the pressure tubes broke open, the disruption of the fuel channel tubes (initially the result of power spiking from the positive void coefficient) led to generation of steam in massive quantities from the reactor coolant loop decompression. In addition, the full impact of the intrinsically positive reactor void coefficient became apparent.[5]

The kinetic energy released from the accident sequence has been estimated to be equivalent to 30–40 tons of trinitrotoluene (TNT). Much of the reactor building structure was destroyed in the explosion exposing the reactor core directly to the environment. The explosion sent a plume of thermally hot and highly radioactive particles, smoke, and building debris into the air.

After the explosion, fires began in what remained of the Unit 4 building, in the adjacent turbine hall roof, and in various stores of diesel fuel and flammable materials. In total, more than 35 separate fires started on the plant site as a result of the initial explosion and ejection of hot debris from the reactor core. These fires were put out early on the morning of 26 April. However, by the time these fires were extinguished, the graphite moderator in the reactor core had caught on fire. Because of the high purity of the fuel (essentially pure carbon) and the high temperature of combustion, there was very little if any visible smoke from the fire. There is very little experience—anywhere in the world—fighting graphite fires involving high radiation and radioactive contamination levels associated with the dispersal of nuclear materials from the core. Because of the location of the fire in relation to the exposed core and an ongoing concern about a possible restart of the nuclear chain reaction, the decision was made to smother the fire and hopefully stop any further radioactive particulate releases. Using helicopters, large quantities of boron carbide, dolomite, lead, sand, and clay were flown over Unit 4 and released. Unfortunately, because of the conditions at the time, many of the

---

[5]Editor's Note: The authors do not mention the high-temperature reaction between the zirconium-alloy fuel cladding and steam or water in the accident sequence. Previous accounts by survivors of the accident describe a possible second explosion 3–5 s after the first. Many analysts believe that such an explosion resulted from hydrogen produced by the water-zirconium reaction.

loads were not actually dropped on the fire, but on areas surrounding it [NEA 1995].

Late on 26 April, officials realized that the town of Pripyat would be severely contaminated by the contamination released as a result of the graphite fire and decided to evacuate the town. Arrangements were made for transporting and housing the residents, and the evacuation began at 14:00 on 27 April, more than 36 hours after the start of the accident. As the extent of the radionuclide contamination became clearer, all of the towns within 30 km of the reactor were evacuated [NEA 1995]. In total, about 116,000 people were evacuated. Hot spots of contamination occurred outside of the 30-km zone. In 1990, residents were evacuated from specific hot spots defined in the Ukrainian Law on the Status and Social Protection of Citizen Victims of the Chornobyl Disaster in 1991 (see Table 6.2.5 and Figure 6.2.9 for definitions of the hot spots).

By 9 May 1986, the graphite fire was extinguished. The explosion and subsequent fires released approximately 4% of the nuclear material in the reactor core directly to the atmosphere. On the same day, work began on placing a massive reinforced concrete slab with a built-in cooling system under the destroyed Unit 4 reactor core. This would allow core cooling if necessary and act as a barrier between melted radioactive material and groundwater [NEA 1995].

In mid-May 1986, the Government Commission decided to build a long-term enclosure around Unit 4 to prevent radionuclides from spreading into the environment and to reduce the effects of operational exposure to high-energy radiation at the Chornobyl NPP site. Construction of the Shelter began May 1986 and was completed in November 1986 (see Chapter 3 for more information on the Shelter).

### 2.3.2 Causes of the Accident

The main reason the accident occurred was excessive reliance on administrative controls (including improperly written operating procedures) and human performance rather than engineered safety features and automatic control systems. This was compounded by technical difficulties with the RBMK reactor design, which could not handle the accident or many other accidents, including design-basis accidents [Adamov et al. 1986] and an underestimation of the effect of steam content on RBMK core reactivity. Another source of errors is the small amount of scientific and engineering work performed to validate the physics of neutron processes in RBMK reactor cores and the lack of experimental research under nearest-to-natural conditions. There has never been any fundamental disagreement between various experts with respect to the true causes of the accident.

An analytical approach based on identifying the contradictions between various accident scenarios and the available objective data identified using the DREG (diagnostic registration) software package enabled the Kurchatov Institute in early summer 1986 to identify one potential technical cause from the following list of 13 options that was consistent with all objective data. (The cause was Item 6—effect of displacement units mounted on RCPS control rods):

1. Hydrogen explosion in pressure suppression pool

2. Hydrogen explosion in lower RCPS coolant tank

3. Sabotage (detonation of an explosive charge that destroyed the primary reactor coolant loop)

4. Explosion of the main reactor coolant pump (RCP) high-pressure header or the group distribution header

5. Explosion of the drum-type steam separator or steam-and-water lines

6. Effect of displacement units mounted on RCPS control rods

7. Failure of automatic reactor power control system

8. Gross operator error when controlling manual control rods

9. RCP cavitation, leading to steam and water in the fuel channel

10. Cavitation on throttling control valve

11. Trapping of steam-separation drum steam in downrun lines

12. Steam-zirconium reaction followed by hydrogen explosion in reactor core

13. Compressed gas from cylinders into the emergency core cooling system.

## 2.4 ROOT CAUSES OF ACCIDENT

The most important lesson of the Chornobyl NPP accident is not just that the safety-related physical and engineering infrastructure at nuclear facilities is in need of improvement, but that the basic principles and requirements of a safety culture should be observed in all aspects of atomic energy utilization.

The expression "safety culture" refers not only to the defense in depth design but also to the personal responsibility of all of the staff involved in safety-related activities. Implementation of a safety culture means, among other things, that personnel training should place primary emphasis on understanding the reasons behind safety practices as well as the safety con-sequences for failure to adequately comply with personal obligations. Safety culture presumes an overall psychological bias toward safety, as established by organizational managers involved in NPP construction and operation.

The concept of a safety culture goes beyond the framework of pure oper-ational activities and encompasses any type of activity that might affect safe NPP operation during any phase of the plant's life cycle. The safety culture should permeate upper management, including legislative and governmental management. Under the safety culture concept, legislative and governmen-tal management is charged with the creation of a national climate in which safety is an everyday issue. The safety culture also requires that those with responsibility develop and maintain a questioning attitude—a concept that is very alien in the hierarchical nuclear establishment where the decisions and pronouncement of design institutes and high officials are accepted without the type of substantiation and documentation required in the West.

The safety culture during other Chornobyl NPP construction and opera-tion phases was just as inadequate as during the operations phase, according to an evaluation of the events that unfolded during the accident. An institu-tional framework was not in place to regulate the legal, economic, and sociopolitical relationships for the nuclear power industry in conformance with internationally accepted basic safety standards. There was no single person or organization with authority and responsibility for NPP safety. Organizations involved in NPP operation as well as construction were only responsible for the work that they had actually performed. Various govern-ment ministries were the only entities with high-level decision-making authority over NPPs. Thus, decision-making authority was essentially divorced from any responsibility for the consequences of the decisions.

Generally accepted international practice dictates that the operating organization always bears final responsibility for the safe operation of an NPP. However, this responsibility is meaningless without additional author-ity. Moreover, the former Soviet system did not grant appropriate authority to the NPP management or even the organizations with jurisdiction over the NPPs, which jointly acted as an operating organization. Under the rules and regulations in force at the time, NPP management did not have the authority to make any decisions without the approval of the Lead Design Engineer, Scientific Manager, Lead Project Design Engineer, and the governmental agency responsible for compliance with safety standards. None of these organizations—which essentially issued orders to the nuclear plants—bore any responsibility for the consequences of their decisions.

As early as 1979, the Committee of State Security of the Union of Soviet Socialist Republics (i.e., the KGB) documented significant deviations from design requirements and norms in the construction of Phase II of the

Chornobyl NPP. While the deviations were documented and reported to the Central Committee of the Communist Party of the Soviet Union, no corrective actions were taken because there was no single organization with over-arching responsibility for nuclear safety oversight.

The situation is now quite different: the primary objective in the field of nuclear safety now is learning to live within the law. In February 1995, Ukraine enacted the Law on Atomic Energy Utilization and Radiation Safety. This law embodies all of the generally recognized international principles for safe utilization of atomic energy, with the following primary features:

- Operating organizations have complete and final responsibility for the safety of nuclear installations.

- The regulatory agency is independent of all organizations and individuals responsible for the development of nuclear power; its decisions are final and can only be overruled by court decision, i.e., in the manner prescribed by law.

- Safety has priority over all other goals in atomic energy use.

To date, nuclear safety has not been of prime importance or a psychological need to those responsible for basic decision making, either at the governmental level or at the enterprise or organizational level. The environment in the regulatory arena is becoming difficult—a clear lack of human resources and an almost total lack of material resources have led to two areas of legal activity. One area involves the law and is based on licensed regulation of atomic energy; the other involves the rules and regulations governing nuclear safety, which have remained unchanged and are based on the principles of oversight. Daily activities are conducted on the basis of rules and regulations, so it is not surprising that most experts and managers continue to live and work according to principles inherited from the former Soviet Union.

## 2.5  PHYSICAL AND CHEMICAL CHARACTERISTICS OF RELEASE

To understand the impact of the accident of the Ukraine, Belarus, Russia, and the world, the release—which occurred over several days—of radioactive materials needs to be characterized. Also, the physical and chemical characteristics of the fuel particles collected in 1986–1987, combined with data on the loss of fission products from the particles, might help in understanding the processes that took place in the reactor during the accident. Further, the information might help in estimating the amounts of radioactive inert gases and iodine lost.

The long-term dynamics (over a 10-day period) of the release (Figures 2.5.1 and 2.5.2), coupled with variations in meteorological conditions, led to a complex pattern of fallout over a wide area. According to various estimates [IAEA 1996b], up to 100% of the radioactive inert gases, 20–50% of the iodine isotopes, 12–30% of the $^{134,137}$Cs, and 3–4% of the less volatile radionuclides ($^{95}$Zr, $^{99}$Mo, $^{89,90}$Sr, $^{103,106}$Ru, $^{141,144}$Ce, $^{154,155}$Eu, $^{238-241}$Pu, etc.) in the reactor at the time of the accident were released into the atmosphere. Table 2.5.1 lists the activity of significant radionuclides in the Unit 4 reactor at the time of explosion, along with the estimated amount of radioactivity released [Ministry for Chornobyl Affairs 1996].

### 2.5.1  Explosion and Fire: Radioactive Materials Fractured and Released

The initial release from the explosion on 26 April 1986 led to the deposition of a very narrow (up to 80 km in length and 1–5 km in width) fuel plume toward the west, that is, toward the village of Tolstyi Les. The graphite fire, which burned for about 12 days following the explosion, resulted in additional plumes of radionuclides which fell in detectable amounts throughout Eastern Europe and Scandinavia and were detected as far away as North America. The initial fuel plume generally consisted of finely dispersed fuel particles from the original core disintegration and explosion. This plume

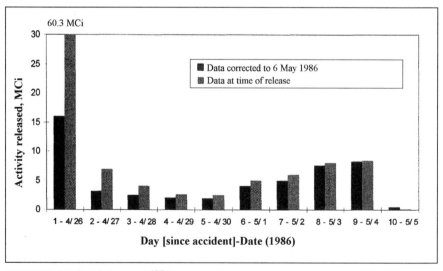

**FIGURE 2.5.1.** Release of $^{137}$Cs from the Chornobyl Unit 4 reactor as a function of time in 1986 [Abagyan et al. 1986; Izrael et al. 1987]

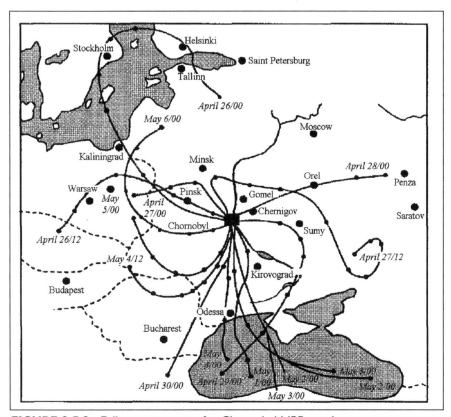

**FIGURE 2.5.2.** Fallout trajectories for Chornobyl NPP accident

contained approximately 10% of the off-site fuel particles ultimately deposited offsite. The cause for the plume was a short-lived, low-altitude (maximum 300–500 m) release of fuel particles, probably accompanied by superheated steam, at night in a stable atmosphere. At the time of the accident, the surface wind was weak and indeterminate in direction, with an 8–10 m/s southeast wind only setting in at 1,500 m altitude [IAEA 1992].

The Chornobyl NPP accident produced hot fuel particle fallout consisting of various impurities in a uranium-oxide matrix. The radionuclide composition of the hot fuel particles was similar to that of the Unit 4 spent fuel at the time of the accident (Table 2.5.1), except for fractionation of highly mobile volatile fission products [Loshchilov et al. 1991; Kuriny et al. 1993; Loshchilov et al. 1993a; Loshchilov et al. 1993b; Kashparov et al. 1997b]. Data on the loss of $^{95}$Zr and $^{141,144}$Ce indicate that approximately 3% of the

**TABLE 2.5.1.** Composition of Radionuclides in Reactor Core and Release at Time of Accident

| | **Content of Reactor Core as of 26 April 1986** | | **Total Release, Recalculated as of 26 April 1986** [Ministry Chornobyl Affairs 1996] | |
|---|---|---|---|---|
| **Nuclide** | **Half-life** | **Activity (x $10^{15}$ Bq)** | **Percent of Total** | **Activity (x $10^{15}$ Bq)** |
| $^{133}$Xe | 5.3 d | 6,500 | 100 | 6,290 |
| $^{131}$I | 8.0 d | 3,200 | 20 | 1,650 |
| $^{134}$Cs | 2.0 yr | 180 | 20 | 52 |
| $^{137}$Cs | 30.0 yr | 280 | 13 | 85 |
| $^{132}$Te | 78.0 h | 2,700 | 25–60 | ~1,020 |
| $^{89}$Sr | 52.0 d | 2,300 | 4–6 | 93 |
| $^{90}$Sr | 28.0 yr | 200 | 4–6 | 8.1 |
| $^{140}$Ba | 12.8 d | 4,800 | 4–6 | 180 |
| $^{95}$Zr | 64 d | 5,600 | 3.2 | 155 |
| $^{99}$Mo | 67.0 h | 4,800 | >3.5 | — |
| $^{103}$Ru | 39.6 d | 4,800 | 2.9 | 170 |
| $^{106}$Ru | 1.0 yr | 2,100 | 2.9 | 59 |
| $^{141}$Ce | 33.0 d | 5,600 | 2.3 | 190 |
| $^{144}$Ce | 285.0 d | 3,300 | 2.8 | 137 |
| $^{239}$Np | 2.4 d | 27,000 | 3 | 1,440 |
| $^{238}$Pu | 86.0 yr | 1 | 3 | 0.03 |
| $^{239}$Pu | 2,4400 yr | 0.85 | 3 | 0.03 |
| $^{240}$Pu | 6,580.0 yr | 1.2 | 3 | 0.044 |
| $^{241}$Pu | 13.2 yr | 170 | 3 | 5.9 |
| $^{242}$Cm | 163.0 d | 26 | 3.5 | ~0.9 |
| Total | | ~73,559 | | ~10,933 |

Editor's Note: Based on normalizing the activity to the midpoint of the release (done on 6 May 1986), the total activity released was approximately 2,000 x $10^{15}$ Bq. Normalizing to the midpoint essentially eliminates the short-lived noble gases. They contribute little to the overall offsite dose.

nuclear fuel was released from the reactor in the form of hot fuel particles during the accident (only a few hundred kilograms of which were released during the explosion on 26 April; the remainder was released over the next several days as a result of the fire). The only previous observation of such particles came from the 1957 Windscale accident in which an air-cooled graphic reactor caught fire and burned for several days [Salbu et al. 1994].

To understand how the fission products in Unit 4 fractionated, the released fuel particles were studied. The particles can be categorized by relative size, that is, large and small particles. While the large particles have little radiological significance outside the 2- to 5-km zone around the reactor—as they make up an insignificant part of the overall radioactive surface contamination, they may provide insight into the accident. The large fuel particles released during the 26 April 1986 explosion varied in size from several dozen to several hundred microns. Small fuel particles (median radius on the order of 3 $\mu$m) consisting of nonoxidized uranium dioxide crystals (or grains) were also released.

The large particles were depleted of cesium isotopes. This cesium loss can be explained by radionuclide migration within the hot fuel particles and loss of some fission products from the fuel during the high-temperature annealing process in the reactor. The loss of cesium before the accident through normal reactor operations was insignificant, as indicated by the lack of change (taking burnup into account) in the ratio of $^{134}$Cs and $^{137}$Cs activities, which have the more mobile $^{133}$Xe and $^{137}$Xe (with different half-lives) as precursors. Cesuim-134 is produced by neutron activation of stable $^{133}$Cs, the decay product of $^{133}$Xe. Thus, if the fuel had lost cesium during normal reactor operations, the fuel particles would have been more depleted in $^{134}$Cs as a result of xenon migration. No fractionation of $^{144}$Ce, $^{95}$Zr, $^{125}$Sb, $^{154,155}$Eu, or $^{238-240}$Pu, all of which have small migration coefficients in uranium dioxide, was observed in the fuel particles. These radionuclides have relative fractionation coefficients of 1, so a zero loss from the fuel particles can be assumed.

Experimental data on the relative loss rate of $^{137}$Cs from the Chornobyl fuel particles was used to obtain an estimated equivalent spherical radius (uranium dioxide grain or crystallite) of $L = 2.4 \pm 0.7 \mu$m for the Chornobyl nuclear fuel (Table 2.5.2). This value is highly consistent with the observed median radius of 1.5–3.5 $\mu$m (lognormal distribution in the soil) for the fuel component in Chornobyl fallout, as well as the size of the grains into which Chornobyl nuclear fuel crumbles upon oxidation in air (3 $\mu$m) [Kashparov et al. 1997b].

Burnup-corrected data on the fractionation coefficients of $^{137}$Cs and $^{90}$Sr in the Chornobyl fuel particles and an experimentally determined function describing the effective diffusion coefficients as a function of temperature were used to determine the effective annealing time and temperature for the fuel particles during the accident [Kashparov et al. 1997b, Kashparov et al. 1997a]. The resulting effective temperatures, temperature distributions, and distributions of isothermal annealing times for the particles released from the reactor (with median values of $T = 2400$ K and $t = 3.5$ s, respectively) indicate that the fuel particles ($>20\ \mu$m in diameter) were formed "explosively," with the annealing temperature rapidly increasing during the accident. Similar median temperatures and slightly longer median annealing times were also obtained under nonisothermal (explosive) conditions assum-

**TABLE 2.5.2.** Experimental Ratios of the Activity of Radionuclide $i$ ($A_i$) to $^{144}$Ce ($A_{HFP}(^{144}$Ce)) Activity in Hot Fuel Particles at the Time of the Accident

| Radionuclide $i$ | $A_i/A_{HFP}(^{144}\text{Ce})$ |
|---|---|
| $^{90}$Sr | 0.05 |
| $^{95}$Zr | 2.3 |
| $^{103}$Ru | 1.08 |
| $^{106}$Ru | 0.26 |
| $^{125}$Sb | 0.006 |
| $^{134}$Cs | 0.02 |
| $^{137}$Cs | 0.04 |
| $^{141}$Ce | 1.34 |
| $^{144}$Ce | 1.0 |
| $^{154}$Eu | 0.0015 |
| $^{155}$Eu | 0.0017 |
| $^{239,240}$Pu | 0.0004 |

ing a linear or exponential increase in temperature ($T = 2630$ K and $T = 2640$ K, respectively, and $t = 17$ s and $t = 27$ s, respectively).

The large fuel particles (which should not be confused with agglomerates of fine oxidized fuel particles on various substrates) were primarily deposited within 5 km of the reactor. The large particles were formed at high temperatures during the initial phase of the accident as a result of fragmentation of the fuel (sharp increase in power density as a result of the increased neutron flux, shock wave, temperature gradients, etc.), as evidenced by the fact that the fuel burnup within the large particles differs from the mean burnup in the reactor (Figure 2.5.3). Moreover, the ratio of $^{95}$Zr/$^{144}$Ce activities in the large particles is a factor of 1.5 higher than the same ratio in any of the soil samples collected from the fuel plumes beyond the Chornobyl NPP 5-km zone (with the exception of the narrow plume to the west from the initial release) or the mean value for the reactor as a whole.

By matching the fuel burnup to that of the particles, the approximate location of the explosion in the reactor core was determined [Kashparov et al. 1997a]. The determination rested on the following assumptions:

- At the time of the accident, there was substantial variation in the nuclear and physical characteristics of the fuel within Unit 4 because of the different burnup levels for various individual fuel assemblies.

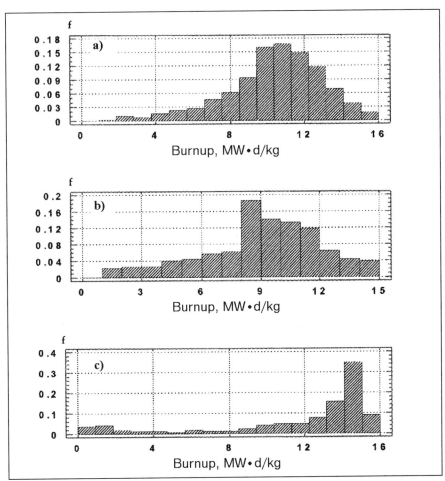

**FIGURE 2.5.3.** Distribution of fuel particles with respect to burnup:
a) recovered particles classed on the basis of $^{137}Cs/^{134}Cs$ ratio; b) recovered particles classed on the basis of $^{125}Sb/^{144}Ce$ ratio; c) baseline Unit 4 core

- The accident converted the nuclear fuel into large (>20 $\mu$m) fuel particles that were released from the area surrounding the initial explosion.

- The nuclear fuel burnup as a function of length along each fuel assembly is assumed to follow a Gaussian distribution peaking at the center of the fuel element; the calculated burnup values adequately reflect the state of the reactor at the time of the accident.

- The explosion occurred within a compact region of the reactor core; the amount of fuel released as a result of the explosion is estimated to be on the order of a few hundred kilograms.

This method of determining the site of the explosion was based on identifying a compact region, consisting of one or more fuel assemblies within the reactor core, such that the nuclear fuel burnup at the time of the accident corresponds to the observed fuel-particle burnup distribution. Maps showing the distribution of burnup throughout the reactor core were generated using computer software developed for this purpose.

Burnup value analysis shows that the initial explosion was located in the lower, southeast section of the reactor core. The analysis rules out the upper and lower surfaces of the reactor core as potential sites of the initial explosion, as the particles released would have had lower burnup values than those determined by the Hot Particles Database. Upon further analysis of the southeastern region, we have identified a region with relatively good agreement between the observed and calculated distribution of nuclear-fuel burnup for the fuel particles (Figure 2.5.4). This region includes 55% of the fuel in the fuel assembly, and is located approximately 75% of the way down the fuel assembly. To determine the area with more accuracy will require additional information—for example, information on pre-accident refueling of the fuel assemblies.

The $^{95}Zr/^{144}Ce$, $^{95}Zr/^{125}Sb$, and $^{95}Zr/^{154,155}Eu$ ratios are 1.5 times higher for the large fuel particles released during the accident than for the reactor core as a whole or for the oxidized fuel component of the finely dispersed fallout. This cannot be explained in terms of activated zirconium from fuel-element cladding structures or by generation of zirconium during the explosion. One explanation is that before the accident (within 2 months or so) fuel assemblies were moved from a region of low neutron flux to a region of high neutron flux.

### 2.5.2 Loss of Radionuclides Following Explosion and Other Processes

Residual heat production and heating of the fuel cluster (or clusters) to 1800–2300 K  (26–29 April 1986) led to the loss of high-mobility volatile fission products (the radioactive inert gases, iodine, tellurium, and cesium). The heat production and cluster heating also led to the formation of a convective plume, more than 1 km high on 26 April 1986 and 600 m high on subsequent days, containing these products [IAEA 1992; Izrael 1990]. The plume led to the formation of radioactive condensation plumes in which the fraction consisting of oxidized fuel particles remained virtually constant. The oxidized particles were formed at temperatures of less than 1200 K at the edge of the heated region.

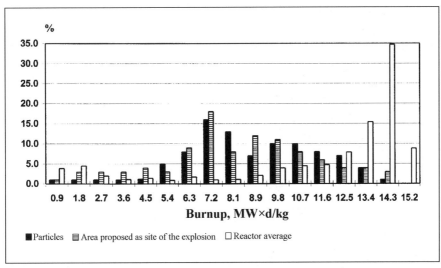

**FIGURE 2.5.4.** Distribution of fuel particles, nuclear fuel in the area proposed as the site of the explosion, and nuclear fuel in the reactor as a whole as a function of burnup

The most plausible timeline for the release of the oxidized fuel particles seems to be discussed in *Chornobyl: Radioactive Contamination of the Environment* [Izrael 1990] and shown in Figure 2.5.1. According to this timeline, the cesium "hot spots" far from the accident site were due to precipitation-related $^{137}$Cs fallout. As the reactor temperature decreased (30 April–3 May 1986), with a corresponding reduction in the rate at which the fuel lost volatile fission products, aerial dispersal from nuclear fuel oxidation became more important. While cesium loss is an increasing function of temperature, and only becomes detectable at temperatures greater than 1500–1800 K, oxidation of uranium dioxide proceeds most rapidly at 600–1200 K, peaking at 800–1000 K. This was what caused the southern fallout plume.

After the reactor was covered with material,[6] this reduced the ability of the fuel to cool, and thus temperatures increased again (3–6 May 1986),

---

[6]Editor's Note: In an attempt to ensure that the reactor was shut down and to limit the further release of radioactive material, approximately 5,000 metric tons of sand, clay, dolomite, and lead was dropped onto the remains of Unit 4 during the week following the accident. The effectiveness of this campaign is in dispute but is believed by most Western experts to be ineffective.

volatile fission products were lost, and materials used to cover the fuel melted. The interaction of the high-temperature core debris with the concrete and sand present in the structure led to the production of aluminosilicate and a variety of eutectic compounds. Aluminosilicates form refractory compounds with many fission products and are highly suited for fixing cesium and strontium. Thus, variations in fuel annealing temperature during the accident played a very important role at various times, both with respect to loss of various volatile fission products (such as radioactive inert gases, cesium, and iodine, which have quite different migration parameters) and with respect to the fraction of fuel particles (Figure 2.5.1).

### 2.5.3 Fuel Disintegration as a Result of Oxidation in Air

When annealed in air for 3 h at 673 K, the actual fuel from Chornobyl NPP disintegrates into fine particles that are similar in size to the uranium dioxide fuel grains. Figure 2.5.5 shows the post-annealing particle-size distribution $f(g)$ as a function of annealing time ($t$ = 3–21 h) and temperature ($T$ = 673 K). These distributions are described quite well by a lognormal distribution for all values of $t$ (3–21 h) and $T$ (673–1173 K) used in the experiments (see Equation 2.5.1):

$$f(r) = \frac{1}{\sqrt{2\pi} \times s \times r} \exp\left[-0.5\left[\frac{\ln(r) - m^2}{s}\right]^2\right] \qquad (2.5.1)$$

where $m$ is the logarithm of the median radius and $s$ is the dispersion.

The parameters of this distribution are tabulated as a function of annealing time and temperature in Table 2.5.3.

Figure 2.5.5 and Table 2.5.3 indicate that the mean particle size decreases with increasing annealing time. The table clearly shows that the parameters of the distribution are virtually independent of annealing temperature, while the mean log radius of the resulting particles (median radius) generally decreases from 10–3 $\mu$m as the annealing time increases. This is apparently due to movement of oxygen into the particle matrices and the resulting disruption of the particles.

The resulting distributions give us an idea of the particle sizes that result from annealing of actual Chornobyl nuclear fuel, at least for temperatures 673 K < $T$ < 1173 K and annealing times 4 h < $t$ < 16 h.

These distributions are quite consistent with actual lognormal distributions from Chornobyl-related hot fuel particles in the soil. The soil distributions were determined autoradiographically in 1987 and 1989 at various distances from the reactor (5, 8.3, 10, 14, and 30 km) at 10-degree intervals. In

**TABLE 2.5.3.** Parameters of the Fuel-Particle Size Distribution and Median Radius ($R_m$, $\mu$m) as a Function of Fuel Annealing Time and Temperature

| Annealing time (h) | T = 673 K | | | T = 873 K | | | T = 1173 K | | |
|---|---|---|---|---|---|---|---|---|---|
| | m | s | $R_m$ | m | s | $R_m$ | m | s | $R_m$ |
| 3 | 2.06 | 0.60 | 7.85 | | | | | | |
| 4 | | | | 2.25 | 0.81 | 9.49 | 2.29 | 0.58 | 9.87 |
| 7 | 1.49 | 0.77 | 4.44 | | | | | | |
| 8 | | | | 1.41 | 0.53 | 4.09 | 1.31 | 0.45 | 3.71 |
| 12 | | | | 1.47 | 0.53 | 4.35 | 1.38 | 0.44 | 3.97 |
| 13 | 1.35 | 0.54 | 3.86 | | | | | | |
| 16 | | | | 1.16 | 0.56 | 3.19 | 1.12 | 0.47 | 3.06 |
| 21 | 1.09 | 0.36 | 2.97 | | | | | | |

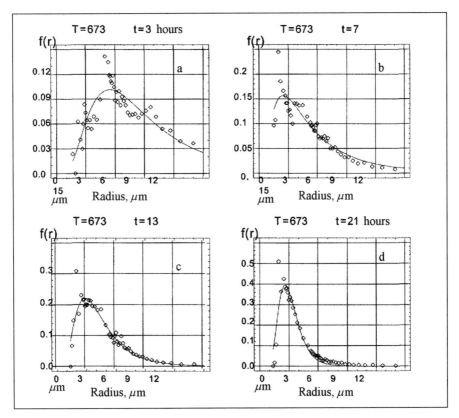

**FIGURE 2.5.5.** Normalized distributions for radii of particles formed after annealing a sample of Chornobyl nuclear fuel at 673 K for a) 3 h; b) 7 h, c) 13 h, and d) 21 hof time in 1986.

1987, the hot fuel particles in the soil had a median radius of 1.5–2.5 $\mu$m; by 1989, the median radius was 2.0–3.5 $\mu$m. The decrease in the percentage of small particles as a function of time is because the particles oxidize easier and are less stable in soil. The percentage of large particles decreases with distance from the reactor; this is because the large particles have a higher dry deposition rate. As gravitational deposition is not the primary deposition mechanism for small particles (i.e., those smaller than 5 $\mu$m), and the deposition rate for such particles is not a strong function of particle size, the fuel-particle fallout size distribution is virtually independent of distance for 5–100 km from Chornobyl NPP (Figure 2.5.6).

The Chornobyl fallout also includes ruthenium particles consisting largely of [103]Ru and [106]Ru in an iron-group matrix. Such particles are rarer than the fuel particles, with ruthenium particles in the 60-km zone accounting for only a few percent of the number of fuel particles. These particles may have been formed through oxidation of ruthenium on the surfaces of dispersed nuclear fuel particles and subsequent reduction from the volatile oxide, $RuO_4$, on structural material particles. Such particles could also have formed in the nuclear fuel during normal reactor operations through formation of the transition metals ruthenium, rhodium, and palladium in the metal phase and molybdenum and technetium in the metal phase or oxide phase. Most of the fuel-particle fallout occurred within 60 km of Chornobyl NPP; the fraction of ruthenium particles in the fallout increases with distance from the source; however, the ruthenium particles have little radiological significance relative to cesium condensate.

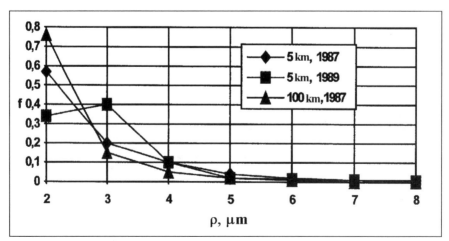

**FIGURE 2.5.6.** Radius distributions in soil for hot fuel particles released in a southerly direction (180°): ◆ 5 km, 1987; ■ 5 km, 1989; ▲ 100 km, 1987. The x-axis reflects the particle radius in $\mu$m. The y-axis is the fraction of the total particles recovered.

The accident also led to the formation of condensate particles. These particles were created through the condensation of highly mobile, volatile fission products (radioisotopes of iodine, tellurium, cesium, strontium, etc.), which escaped from the nuclear-fuel matrix at high temperature onto dust, structural materials, etc. The annealing temperature for the nuclear fuel during the accident determines the loss of such fission products as a function of time. The specific activities of the various radionuclides in the condensate particles are much lower than those in the fuel particles. The formation of the particles is strongly related to condensation time (duration of the condensation process), condensation temperature, and the surface characteristics of the particles, which means that they can vary over a quite wide range. These particles are similar to those formed in nuclear explosions during the final phase when the temperature is relatively low, and as such, the mobility of radionuclides in soil and other media as well as biological accessibility have been extensively studied.

Uranium-dioxide grains (crystallites), also called fuel particles, were also formed by the accident. The grains have a median radius on the order of 2–3 $\mu$m (aerosol median aerodynamic diameter = 10–20 $\mu$m), and can be classified into two types:

- **Unoxidized fuel particles** from the initial release on 26 April 1986; these particles made up the narrow plume to the west (on the order of 10–15% of the off-site fuel-particle activity)

- **Oxidized fuel particles** formed through oxidation of nuclear fuel between 26 April and 5 May 1986; these particles made up the northern (60–70% of the activity) and southern plumes (on the order of 20–25% of the activity).

During the accident, the nuclear fuel came in contact with a wide variety of materials (structural materials: zirconium, graphite, stainless steel; and materials used to bury the reactor such as lead and aluminosilicates). This led to the formation of conglomerates with fine-grained fuel particles and nonradioactive matrices, as well as complex chemical compounds, especially on the surfaces of hot fuel particles (presence of zirconium, silicon, aluminum, lead, etc.). Other researchers have described these conglomerates using the concept of "graphite particles," "structural particles," etc. These conglomerates vary tremendously and contain radiologically significant radionuclides, but they are all based on small-grained particles of nuclear fuel.

Generally, the radiologically significant radionuclides such as $^{90}$Sr, $^{238-241}$Pu, and $^{241}$Am were only released from the reactor during the accident in fuel-particle matrices (more than 90% by activity), which led to the contamination of a large area. Within the 30-km Exclusion Zone (Figure 2.5.7), a significant fraction of the $^{137}$Cs was also initially contained in the fuel particles—more than 75% at distances up to 10 km from Chornobyl NPP and more than 50% at distances up to 30 km in the western and southern plumes.

The radionuclides in the initial fallout, which were encapsulated in a fuel-particle matrix, had lower migration mobilities than the same radionuclides found further from the accident because of different dispersion methods and other factors. Thus, applying migration laws that were originally derived for global fallout to the initial fallout from the Chornobyl NPP accident led to a substantial overestimation of the radionuclide migration rates for various food chains in the Exclusion Zone.

### 2.5.4 Behavior of the Fuel Component of Fallout

The fuel particles eventually dissolved, and the radionuclides leached out of the fuel particles, as indicated by the changes in the fuel-particle size

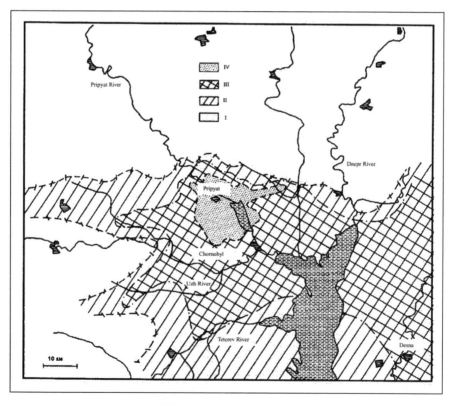

**FIGURE 2.5.7.** Fraction of $^{137}$Cs in initial fallout in fuel particles (R) (I R < 25%; II 25% < R < 50%; III 50% < R < 75%; IV R > 75%)

distribution and the elevated level of metabolizable strontium in the soil, and, thus, increased contamination of vegetation. It was not possible to predict or model the behaviors with any reasonable degree of accuracy. Thus, derivation of mathematical functions for the fuel-particle dissolution rate and rate of radionuclide leaching from the fuel particles under a variety of natural soil and climatic conditions would be exceedingly important to developing long-term predictions for the behavior of radioactive materials in the Exclusion Zone, or in the event of another accident involving the release of spent fuel particles.

Analysis of the metabolically available $^{90}$Sr levels and levels of $^{85}$Sr indicator inserted into the soil indicates that for the dissolution rate of fuel particles in soil under natural conditions, the physical and chemical properties of the particles themselves are more important than soil and climatic conditions [Kashparov et al. 1997b]. For example, the particles in the initial, westward (270°, narrow western fuel plume) release, which had the lowest levels of uranium dioxide oxidation, had the highest chemical stability, independent of soil characteristics. Between 27% and 70% of the $^{90}$Sr activity is still contained in the fuel-particle matrices. On the other hand, the particles that were annealed in the reactor for an extended period and were released in other directions after 26 April 1986 dissolved to a greater extent under similar soil conditions. The most important factor affecting the dissolution rate for identical particles is soil pH. The best-preserved particles were in weakly acidic soil (pH > 5.2) along the western plume, especially in peat soils with pH = 5.2–5.5 and derno-podzolic loam with pH = 5.6.

Based on the actual size distribution of the fuel particles in Chornobyl fallout, the measured fraction of undissolved particles, and the change in fuel particle concentration ($\Delta$FP) remaining after 9 years, Equation 2.5.2 was used to calculate the linear dissolution rate V ($\mu$m/yr) for the particles at each experimental point, which was assumed to remain constant in time:

$$\Delta FP = \frac{\int_b^\infty (r-b)^3 f(d)dr}{\int_0^\infty r^3 f(r)dr} \tag{2.5.2}$$

$$b = V \times t$$

where $t$ is the time post-accident over which the particles dissolve in the soil (yr).

The linear dissolution rate for the weakly oxidized fuel particles released toward the west (270°) during the initial explosion and for the highly oxidized particles released toward the north and south as the reactor burned is strongly correlated with soil acidity (pH). The following relationships were obtained for the dissolution rate of the fuel particles as a function of soil acidity:

- For the weakly oxidized fuel particles to the west of Chornobyl NPP,
  $$V = 14.45 \times 10^{-0.431 \times pH} \; \mu m/yr$$

- For the highly oxidized particles to the north and south of Chornobyl NPP,
  $$V = 4.59 \times 10^{-0.234 \times pH} \; \mu m/yr$$

The resulting function (Figure 2.5.8) describes the dissolution rate for fuel particles in typical Exclusion Zone soils at various moisture levels for $4 < pH < 6$. This function enables us to predict the amount of radionuclide leaching from the fuel particles as a function of time, i.e., the relative number of particles that dissolve under various soil conditions for various directions of release (Figure 2.5.8).

The meadow-vegetation transfer coefficient for strontium and the contamination of vegetation as a function of time (Figure 2.5.9) will each vary in a similar fashion. The resulting predictions are highly consistent with data on $^{90}Sr$ forms in the soil within the Chornobyl plumes. For example, the ratio of the strontium and cesium grass transfer coefficients in the western and southern plumes differed by a factor of 10 because of the different fuel-particle dissolution rates. Thus, a significant fraction of the radioactive material in the narrow fuel plume to the west of Chornobyl NPP (up to 70% at $pH = 6$) will remain in the particles even into the year 2000 (Figure 2.5.8), while most of the fuel particles at other locations will have already dissolved.

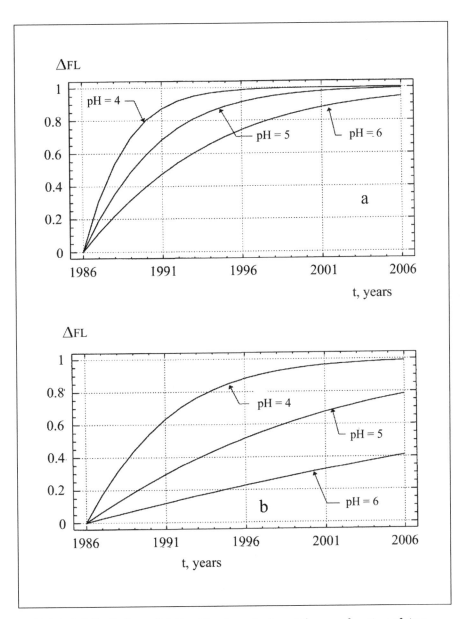

**FIGURE 2.5.8.** Radionuclide leaching from fuel particles as a function of time for various soil acidities and directions of release (a = north and south; b = west), as calculated using the linear dissolution rates (V, $\mu$m/yr) for the fuel particles in soil. $\Delta$FL is the relative number of particles dissolved, $\Delta$FL = 1 − $\Delta$FP.

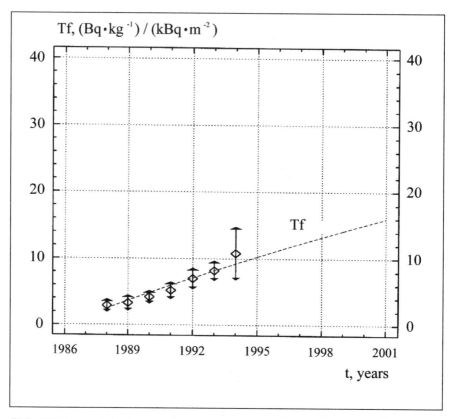

**FIGURE 2.5.9.** Effective meadow-vegetation transfer coefficients for $^{90}$Sr ($T_f$) neglecting redistribution in root layer (data points = experimental values; theoretical curve for pH = 6)

# 3
## укрытие

# THE SHELTER
## *Containing the Destroyed Reactor*

V. Bar'yakhtar, V. Poyarkov, V. Kholosha, and V. Kukhar'

This chapter describes construction of the structure (known as the Shelter) covering the remains of the destroyed reactor, meteorological and geological conditions at the Chornobyl NPP site and how they relate to the Shelter, physical conditions within the Shelter and structural concerns, locations and amounts of radioactive materials in and around the Shelter, a nuclear safety assessment of the Shelter, and analyses of the consequences of collapse of the Shelter roof.

## 3.1 PURPOSE AND GENERAL DESCRIPTION OF THE SHELTER

The reactor explosion at Chornobyl NPP Unit 4 destroyed major portions of the reactor unit building, deaerator mezzanine, turbine hall, and nearby support structures. The primary building damage that could be identified from an external inspection and by entering accessible facilities (limited by radiation levels and physical damage) is summarized here from early first-hand accounts [Kurnosov et al. 1988; Ukrainian Academy of Sciences 1992; Borovoi 1990].

- ***Reactor Unit Damage:*** The reactor core was entirely destroyed. Fragments, including fuel rod fragments, graphite core bricks, and structural elements, were propelled by the explosion into the building rubble, onto the roofs of neighboring buildings, and onto the ventilation stack, and were scattered over the adjacent territory. Later, scientists determined that a portion of the nuclear fuel ended up in the lower elevation of the reactor building as fuel lava. The ventilation stack sustained significant structural damage from the accident.

The upper biological shield slab (System E), which weighed 1,000 metric tons, was forced 14 m into the air and ended up wedged at a 30-degree angle in the top of the reactor shaft. The walls and ceilings over the central reactor hall were destroyed. The walls on the steam-separator buildings were destroyed, and the ceilings were displaced. The refueling machine broke away and collapsed. The northern reactor coolant pump room completely collapsed and was filled with building rubble, while the southern reactor coolant pumps were partially destroyed.

- *Deaerator Mezzanine:* The upper two floors were destroyed and the frame columns were shifted more than 15 degrees from vertical toward the turbine hall.

- *Turbine Hall:* The turbine hall roof collapsed as a result of the fire and the weight of the building wreckage that landed on it. The shock wave from the explosion deformed several structural trusses and displaced the frame columns.

- *Reactor Building Auxiliary Systems:* Only local damage was sustained. Some of the peripheral rooms in the lower elevations of the reactor building showed little evidence of damage from the reactor explosion.

- *Emergency Core Cooling System:* The core cooling system was destroyed in the explosion, which severed all piping attached to the reactor. Major portions of the system were buried under the wreckage of the structure near the reactor.

### 3.1.1 History of Shelter Construction

In mid-May 1986, the Government Commission decided in favor of long-term enclosure of Unit 4 to prevent radionuclides from spreading into the environment and to reduce the effects of operational exposure to high-energy radiation at the remainder of the Chornobyl NPP site.

The Central Committee of the Soviet Union Communist Party and USSR Council of Ministers issued Decree No. 634-188 (dated 29 May 1986). This decree assigned the task of "constructing an enclosure for Chornobyl NPP Unit 4 and associated structures" to the Ministry of Mid-Level Industrial Engineering (MinSredMash), the ministry responsible for the nuclear power industry in the Soviet Union. This enclosure is now called the Chornobyl NPP Unit 4 Shelter or simply the Shelter. Because of the importance of the structure, 18 different designs were generated. These designs can be divided into two groups: the first group included proposals for the construction of a large standalone airtight structure—either an arch with a span of 230 m or a series of domes with a maximum span of 120 m. The second group of designs used the remnants of the destroyed unit to the maximum possible extent. In these designs, the new walls and roof of the proposed Shelter were supported by the remains of the destroyed unit. The second approach was chosen (Figure 3.1.1).

Shelter construction was overseen by two groups. The Kurchatov Institute of Atomic Energy was assigned scientific oversight of the work. A Chornobyl NPP Headquarters Organization was established (along with a special general-contractor organization—Construction Administration 605) within MinSredMash for management and oversight of all work to contain the wreckage of Unit 4.

Construction of the Shelter began in May 1986 and was completed in November 1986. On 30 November 1986, a State Acceptance Commission appointed under USSR Council of Ministers Decree No. 2126rs (dated 23 October 1986) accepted the conserved Chornobyl NPP Unit 4 for maintenance and surveillance.

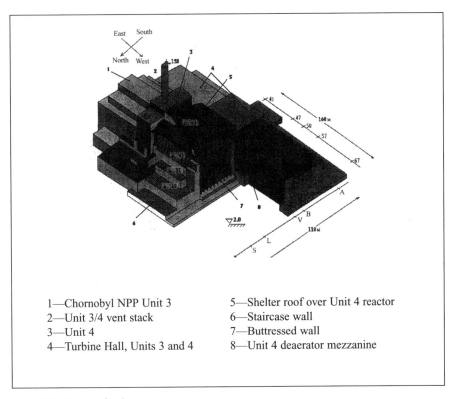

1—Chornobyl NPP Unit 3  
2—Unit 3/4 vent stack  
3—Unit 4  
4—Turbine Hall, Units 3 and 4  
5—Shelter roof over Unit 4 reactor  
6—Staircase wall  
7—Buttressed wall  
8—Unit 4 deaerator mezzanine  

**FIGURE 3.1.1.** Shelter

### 3.1.2 **Primary Construction Stages of Shelter**

The All-Union Scientific Research and Design Institute for Electrotechnology in St. Petersburg was the lead design organization for the Shelter. Other organizations involved in designing various individual structures were the Leningrad Structural Element Design Institute, St. Petersburg; the Central Scientific Research Institute for Structural Element Design, Moscow; the Ukrainian Scientific Research Institute for Structural Element Design, Kiev; and the Dnepropetrovsk Institute for Structural Element Design, Dnepropetrovsk [Kurnosov et al. 1988; Borovoi 1990, Ukrainian Academy of Sciences 1992; Borovoi et al. 1990b].

The first step in constructing the Shelter was to erect walls and partitions to separate damaged Unit 4 from Unit 3 and the equipment shared by both units, particularly the Ventilation Building (Block "V"). The protective "construction walls" were intended to support safe operations during Shelter construction (Figure 3.1.2) The walls—made of reinforced concrete—were constructed around the perimeter of Unit 4 to the following heights:

- Approximately 6 m along the collapsed northern side of the unit

- Approximately 8 m along the southern and western sides.

These walls provided the foundation for the buttress walls of the Shelter. The stepped (also referred to as cascade or staircase) wall along the north side of the unit consisted of several concrete "steps" approximately 12 m high (Item 6 in Figure 3.1.1). Metal plates were used as forms. Each successive step was built as close to the destroyed unit as possible. The steps also contain worn-out and damaged structural elements and containers of high-level radioactive waste. The remaining wall of the unit (the western wall) was covered by a wall with buttresses up to 50 m high.

An enormous amount of effort went into decontaminating the surrounding area during construction of the Shelter [Borovoi 1990; Panfilov et al. 1989; Belyaev and Borovoi 1991]. The radiation field in the surrounding area decreased as the decontamination effort continued and the Shelter walls were constructed. By the time construction was complete, the exposure rate in the surrounding area had decreased to 5–10 mGy/h (0.5–1 R/h). Systematic measurements in 1987 showed a further decrease in dose. Decontamination of the Shelter site and roof led to a rapid decrease in the exposure rate outside the facility.

### 3.1.2.1 **Advantages and Disadvantages to Shelter Construction Method.** In looking at the construction, it is critical to understand the basic advantages and disadvantages of the design selected for the Shelter. The selected approach led to a substantial savings in terms of cost and speed of

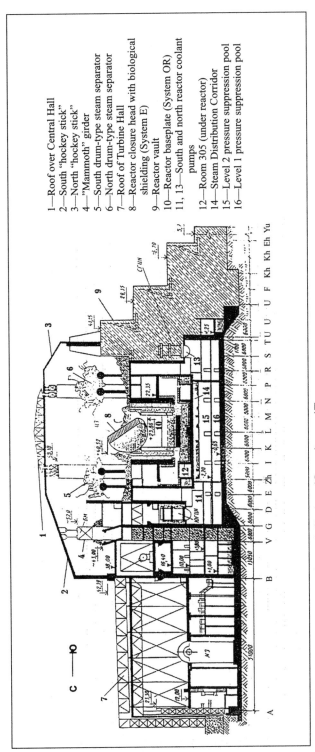

1—Roof over Central Hall
2—South "hockey stick"
3—North "hockey stick"
4—"Mammoth" girder
5—South drum-type steam separator
6—North drum-type steam separator
7—Roof of Turbine Hall
8—Reactor closure head with biological shielding (System E)
9—Reactor vault
10—Reactor baseplate (System OR)
11, 13—South and north reactor coolant pumps
12—Room 305 (under reactor)
14—Steam Distribution Corridor
15—Level 2 pressure suppression pool
16—Level 1 pressure suppression pool

**FIGURE 3.1.2.** Cross-section through Shelter (Coordinate 47)

construction. Indeed, the period between the decision to construct the Shelter and its completion was only half a year—unprecedented for a structure of this size and complexity.

However, this gain came at a price, both in terms of the enormous collective dose[1] received by the construction workers and the fundamental limitations of the facility itself. The need to erect new structures in the immediate vicinity of the destroyed unit (with enormous radiation fields) forced the use of remote-control technology—demag cranes, Putzmeister remote-control concrete pumps, etc. Welds could not be used for many of the load-bearing structures, and remote installation did not allow precise relative placement of the large metal structures used. Concrete flowed into the damaged building, hindering (or even completely preventing) access to and inspection of many locations. As a result, the Shelter is not airtight and has a large number of cracks and openings. (The total area for these openings was subsequently estimated to be 1,000 m$^2$ [Borovoi et al. 1990b]). In a concerted effort from 1995–1997, a large number of these cracks and openings were eliminated.

Further, the radiation field and debris hindered a full evaluation of the strength of many of the supporting members. These supporting members (the main load-bearing beams for the entire new structure) were remnants of the original structures and were affected by the explosion and the fire. The condition of the main load-bearing structural elements of the Shelter (both old and new) is discussed in Section 3.2.

3.1.2.2 **Components of Shelter.** Before we continue with our description of the Shelter, we will define the boundaries of the Shelter and its components. The following structures are assigned to the Shelter [Anonymous 1994]:

- Unit 4 of Chornobyl NPP Phase II

- All new structures installed around the destroyed unit since the accident, including the roof over the reactor hall, the collapsed area around the reactor, the deaerator mezzanine, and the turbine hall

- The portion of Chornobyl NPP Phase II shared structure (e.g., Block "V") inside the partition wall

- The portion of the Reactor Auxiliary Systems Building between coordinates U and Yu and 41 and 50

---

[1]Editor's note: The collective dose received in 1986 by 117,000 workers involved in the construction of the Shelter was approximately 9,800 person-Sv [Kryuchkov and Nossovskii 1996].

- The portion of Unit 4 between coordinates A and B from the partition wall to coordinate 68 along coordinate axis 34

- The portion of System E between coordinates B and V from coordinate 41-68

- The area inside the perimeter of the security zone, including the monitoring and surveillance equipment

- All systems intended for ensuring safe operation of the Shelter in normal and emergency modes.

The legal description of the Shelter is based on the *Shelter Facility Charter*. Under Ukrainian State Committee for Oversight of Nuclear and Radiation Safety Decree No. 16 (dated 12 May 1993), the Chornobyl NPP Production Association has temporary authority to act as an "operating organization" as defined in the *General Safety Provisions for Nuclear Plants*, PN AE G-1-011-89 (OPB) [Atimizdat 1976]. On the basis of this authority, the Chornobyl NPP Production Association developed the *Shelter Facility Charter*, which was approved by the General Director of the Shelter on 05 September 1994.

### 3.1.3 **Description of the Radiation Conditions at the Shelter**

3.1.3.1 **Causes of Contamination.** The radioactive contamination of the facility locations and equipment within the facility was caused by the following:

- *Ejection of core fragments* (onto the central hall, the roof of the reactor building and the roofs of neighboring buildings, areas adjacent to the reactor vault, etc.). In 1990, the dose exposure rate was in the thousands of roentgens per hour near the core fragments.

- *Formation and flow of lava-like fuel-containing material* into various locations under the reactor. The dose exposure rate from this material ranges from hundreds to thousands of roentgens per hour.

- *Contamination by reactor combustion products and gas/dust emission.*

- *Deposition of radioactive aerosols and dust* resulting from air circulation and mechanical transport of the contaminants. This was responsible for contamination of rooms that maintained relative integrity but were interconnected with other rooms through the ventilation systems.

- *Spread of activity by water* (coolant from the damaged loop, firefighting water, rain, etc.). As the water flowed through the rooms, it came into contact with dispersed nuclear fuel and graphite and became an actively mobile source of radioactive contamination. As the radioactive water saturated various concrete structures, it caused these structures to be contaminated to significant depths. Utility corridors, stairwells, and various locations at lower elevations became contaminated.

3.1.3.2 **Dose Exposure Rates Inside and Along the Roofs of the Shelter.** The average dose exposure rate in most lower elevation rooms within the reactor building does not exceed 1 R/h (1996). Exceptions include the pressure suppression pool locations, the steam distribution corridor, and other areas with fuel-containing material.

The surface gamma-ray dose exposure rate for the debris piles, which consist of structural members and materials dropped into the Central Hall from helicopters in an effort to smother the graphite core fire, has a significant spread ranging from hundreds of milligrays per hour to several hundred grays per hour (adjacent to exposed reactor core fragments).

The rooms in Unit 4 were classified in terms of dose exposure rate based on results from work performed between 1988 and 1991 by the All-Union Scientific Research and Design Institute for Electrotechnology [Morozov et al. 1990]. The dose exposure rate measurements were conducted using IKS-D thermoluminescent dosimeters for dose fields above 0.3 Gy/h (30 R/h) and using DP-5V, DRG-01 and DKS-01 thermoluminescent dosimeters for dose exposure rates less than 0.3 Gy/h (30 R/h) (Table 3.1.1).

3.1.3.3 **Airborne Contamination in the Shelter.** In the Shelter, airborne contamination was analyzed rather extensively where drilling sets and extracted active cores were examined as well as where auxiliary equipment was located (Table 3.1.2) [Bondar' et al. 1991]. In the active drilling phase (1988–1990), part of the testing done in the area, the number of personnel periodically located in these work areas reached 100. Alpha aerosols presented the primary hazard to operating personnel.[2]

Following the termination of active operations at the facility (1992–1995), including termination of drilling, the aerosol activity level at the decontaminated locations of the Shelter continued to drop. In 1996, the airborne alpha and beta levels did not generally exceed the maximum permissible concentrations (MPC).

---

[2]Editor's note: Refer to the appendix for essential radiological information

**TABLE 3.1.1.** Distribution of Surveyed Rooms/Locations in Unit 4 by Class (Dose Exposure Rate) at Various Building Elevations

| Elevation | Number of Locations Surveyed | Up to 500 $\mu$Gy/h | Up to 1 mGy/h | Up to 10 mGy/h | Up to 100 mGy/h | Up to 300 mGy/h | <300 mGy/h |
|---|---|---|---|---|---|---|---|
| +0.00 | 9 | 3 | — | 2 | — | — | 4 |
| +3.00 | 5 | 1 | — | — | — | — | 4 |
| +6.00 | 32 | 5 | 2 | 12 | 8 | 1 | 4 |
| +9.00 | 17 | 2 | 2 | 1 | 2 | 6 | 4 |
| +12.50 | 16 | 1 | 2 | 5 | 5 | 2 | 1 |
| +19.50 | 14 | 2 | 1 | 4 | 4 | 3 | — |
| +24.00 | 28 | — | 1 | 2 | 16 | 8 | 1 |
| +27.00 | 16 | 2 | — | 2 | 7 | 1 | 4 |
| +31.50 | 13 | 1 | — | 6 | 4 | 1 | 1 |
| +35.50 | 11 | — | — | 2 | 7 | — | 2 |
| +39.50 | 9 | — | — | 1 | 8 | — | — |
| +43.00 | 2 | — | — | — | 1 | 1 | — |
| +49.95 | 5 | — | — | — | 3 | 1 | 1 |
| Total | 177 | 17 | 8 | 37 | 65 | 24 | 26 |

**TABLE 3.1.2.** Impact of Drilling on Airborne Contamination in the Shelter

| Sampling Location (Room) | $\alpha$-Activity of Air (in Bq/m$^3$)* |
|---|---|
| 207/5 (no drilling) | 790 |
| 207/5 (active drilling) | 19,240 |
| 427 (no drilling) | 370 |
| 427 (active drilling) | 5,550 |

*Isotopic mixture not specified.

### 3.1.4    **Water in the Shelter**

Water enters the Shelter through several pathways [Gerasimova et al. 1992]. One source—whose impact grows over time—is water condensate deposited as dew. The dew forms within the cold locations of the Shelter when radionuclide-saturated water and warm moisture-saturated air penetrate the facility. Condensed water was discovered to have a noticeable impact for the first time in the spring of 1993. The primary causes for an increase in water condensation can be attributed to changes that have occurred in the facility. These changes include cooling of the facility because of the decreased decay heat level from the fuel, a reduction in various artificial heat sources from personnel activity in the Shelter, and coverage of interior openings, which prevents rapid heat exchange and ventilation. Dust suppression by means of recycled water sprays also contributed recycled water to the Shelter.[3]

Rain and snow enter the Shelter through cracks and openings in the roof and the walls. The upper estimates for the average annual water penetration from all precipitation produced (before the beginning of efforts to seal the roof in 1994) into the interior locations of Unit 4 are as follows: 3,000 m$^3$ for the reactor building, 1,800 m$^3$ for the deaerator mezzanine and 6,000 m$^3$ for the turbine hall. Sealing efforts have closed a number of holes in the roof of the turbine hall and deaerator mezzanine. Long cracks in inclined sections of the roof running from north to south from the piping enclosure have been sealed off. This has prevented water penetration through these openings but, at the same time, has reduced the natural ventilation of the facility.

Radioactive water collects in lower locations of the unit and then exits by pathways for which no sufficient information yet exists. In 1990, samples were taken of the radioactive water inside the Shelter. The total beta activity due to gamma emitters in the water samples obtained was $2 \times 10^3 - 2 \times 10^8$ Bq/L. Cesium isotopes principally in the dissolved state are the primary contributor to total activity (74–100% of activity). The beta emitter activity of the samples for the $^{90}$Sr is $6 \times 10^2 - 6 \times 10^6$ Bq/L (as of 1993). The uranium isotope concentration in the water samples tested is 5–20,000 $\mu$g/L, with a significant portion in the dissolved state. Plutonium activity does not exceed 3,000 Bq/L. The cesium and uranium concentration in the lowest elevations of the Shelter has grown by more than two orders of magnitude since 1991 (from $1.8 \times 10^5 - 9.2 \times 10^7$ Bq/L for cesium and from 10–4,300 $\mu$g/L for

---

[3]Editor's note: The water used for dust suppression is recycled from water collected in lower levels of the Shelter.

uranium). These observations suggest significant fractalization of the fuel matrix and leaching by water entering the facility.

### 3.1.5 Major Potential Hazards

According to *Strategy for Conversion of the Shelter Facility into an Environmentally Safe System, Decision No. 5 of the Governmental Commission on Comprehensive Resolution of ChNPP Issues* [Governmental Commission 1997], the Shelter poses several potential hazards. The basic potential hazards (risks) are as follows:

- Spread of radioactive dust beyond the Shelter as a result of structural member collapse, earthquake, hurricane-force winds, onset of nuclear criticality, etc.

- Propagation of radioactivity (in the liquid phase) beyond the Shelter in such a way that radioactivity enters the groundwater or tributaries of the Pripyat River

- Emergency exposure of personnel in the event of an accident, either as the result of the some trigger event or as a result of human activity at the Shelter

- Injury to personnel as the result of nonradiological factors (falling structural members, fires, electric current, etc.).

To safely manage these risks, a comprehensive approach to safety-related issues at the Shelter was applied. This approach had three main goals:

- Ensure current safety of the Shelter.

- Ensure long-term safety ("stabilization") of the Shelter.

- Prepare for and perform operations aimed at final conversion of the Shelter to an environmentally safe system.[4]

The approach mandated that for each of these goals a set of measures should be implemented to do the following:

- Evaluate and predict potentially hazardous effects and their consequences.

---

[4]Editor's note: The goal of converting the Shelter into an environmentally safe system is not a final state, but one resembling the concept of interim safe storage applied to retired nuclear facilities in the United States and other Western countries. The anticipated period for interim safe storage is nominally 50 to 100 years.

- Perform accident management.

- Protect against potential adverse consequences of accidents.

- Ensure sufficient logistical, personnel, and financial resources.

### 3.1.6 Monitoring and Control Systems for the Shelter

To ensure the ongoing nuclear and radiological safety of the Shelter, a variety of monitoring and control systems have been installed. These systems were conceived and designed by the scientific and technical organizations responsible for the design of the Shelter. They include [Anonymous 1994] the following:

- A system for monitoring radiation and thermal parameters.

- A system for monitoring the activity and nuclide composition of gas and aerosol emissions from the Shelter through the vent stack (VT-2).

- A system to continuously monitor and record the hydrogen concentration in the air under the reactor vault cover and in the central hall of Unit 4.

- A potassium metaborate feed system. This system increases the subcriticality of the fuel-containing materials in the collapsed area around the Unit 4 reactor through introduction of a neutron-absorbing agent. This system is operated manually.

- A dust suppression system to reduce the aerosol activity in the central area of the Shelter by spraying a water-based dust-suppression compound.

- A Sukhotrub fire-suppression system to feed process water from the Chornobyl NPP Phase 2 (i.e., Units 3 and 4) fire-suppression system fire pumps to the sites of any fires on the roof, inside the turbine hall, inside the deaerator mezzanine, or on the Shelter grounds via fire hoses. (Fire trucks may also be connected to the system.)

- A ventilation system, with a forced-air plenum ventilation system to remove air from above the collapsed area around the reactor and with a filtration station and bypass line. The plenum ventilation system is intended for the following purposes:

  —Removal of heat generated by the fuel-containing materials (if natural ventilation becomes insufficient)

  —Treatment of the exhaust air with aerosol filters (before release into the ventilation duct).

As of April 1999, not all of these systems have completed required testing or been shown to be effective.

## 3.2 STATUS OF SHELTER STRUCTURAL MEMBERS

### 3.2.1 Natural and Technogenic Impacts

To place our description of the condition of the Shelter structural elements in context, we present some brief information on the external effects (both natural and technogenic) to which these structural elements are or could be exposed.

**3.2.1.1 Common (naturally occurring) Climatic Influences.** For many years, researchers have recorded the temperature, wind speed and direction, and snow fall for the area surrounding the Chornobyl NPP. These observations are summarized as follows.

- *Temperature*: The coldest month is January, the hottest July. The mean monthly temperature is -6.6°C in January and +19.1°C in July. The mean diurnal variation in outside air temperature is 5.7°C in January and 11.0°C in July; the diurnal variation has a maximum amplitude of 28.1°C in January and 18.2°C in July. The mean maximum air temperature during the summer (June, July, and August) is 24.1°C, while the mean minimum air temperature (for December, January, and February) is 7.6°C.

- *Wind:* Chornobyl NPP is considered to be in Wind Loading Zone 2, with a standard normal pressure of $W_0 = 0.3$ kPa. The mean annual wind speed is 4.2 m/s. The peak mean monthly wind speeds—up to 5.1 m/s—occur during the winter months. During the summer, the mean monthly wind speed varies from 3.4–3.7 m/s.

  The wind direction in this region is predominantly from the northwest (17%). The maximum 1-min. average wind speed from 1945–1993 was 20 m/s, while the maximum instantaneous wind speed was 25 m/s.

- *Snow:* Chornobyl NPP is in Snow Load Zone 2, with a snow load of $S_0 = 0.7$ kPa. The average snow accumulation in open areas is 8 cm; the maximum average snow accumulation for a winter is 17 cm, and the record snow accumulation is 41 cm. The number of days per year with snow cover is 90–102. The standard depth of seasonal freezing for the sandy and sandy loam soils common to the site is 110 cm.

3.2.1.2 **Extreme Natural Impacts.** Researchers have also recorded the extremes of natural impacts in the area surrounding Chornobyl NPP and the Shelter.

- *Record maximum and minimum temperature:* The maximum temperature range at the Chornobyl NPP is -44.9 °C to +45.2 °C (with a frequency of once every 10,000 years).

- *Hurricane-force winds:* The maximum wind speed (with a frequency of occurrence of once every 10,000 years) is 47.3 m/s.

- *Excess snow fall:* The snow load exceeded the nominal value for this region on five occasions between 1965 and 1970. In 1968, the snow load exceeded the maximum value of 1.4 kPa (with a frequency of occurrence of once every 10,000 years) on one occasion.

- *Tornadoes:* The region is located in a zone of elevated tornado hazard. The design-basis characteristics of a possible tornado are as follows:

  —Probability of a tornado passing through the Exclusion Zone: $12 \times 10^{-3}$/yr

  —Probability of a tornado passing through any given point in the Exclusion Zone: $3 \times 10^{-6}$/yr

  —Peak horizontal velocity of rotational motion in tornado wall: 72 m/s

  —Forward velocity of tornado: 18 m/s

  —Length of tornado path: 18 km

  —Width of tornado path: 290 m

  —Pressure differential between center and edge of funnel cloud: 64 kPa.

- *Precipitation and floods:* Precipitation and floods are a potential danger to the structural elements of the Shelter because of the potential for loss of stability in the foundation soil (due to an increase in groundwater level), nonuniform deformation of foundations, and structural damage to the above-ground portion of the Shelter.

  The maximum precipitation per day is 190 mm. The maximum precipitation within a 20-min period is 72 mm (both events occur with a frequency of once every 10,000 years). The maximum precipitation observed during a 20-min period was recorded on 20 June 1957 and was equal to 31 mm.

  Approximately 50% of the annual flow of the Pripyat River occurs during the spring flood. The maximum sustained flow rate observed in recent years occurred in 1979 and was 6,000 m³/s (water level 110.0 m).

- *Earthquakes:* Seismic hazards for the Shelter include effects due to earthquakes in nearby regions, as well as local earthquakes in the part of Ukraine that is on a platform.

Data from the Ukrainian National Academy of Sciences Institute of Geophysics indicate that the Chornobyl NPP site may experience seismic events as large as magnitude 6. If the artificial alteration of soil conditions as a result of site excavation is taken into account, this means that such an earthquake would be the equivalent of magnitude 7. Most of this increase in magnitude (+1.01) is a resonance effect caused by the high water content of the soil. The final estimated design-basis seismic hazard to the Shelter corresponds to magnitude 6, while the maximum design-basis earthquake corresponds to magnitude 7. A magnitude 7 earthquake is expected once every 10,000 years.

### 3.2.1.3 Engineering and Geological Conditions at the Shelter Site. A
large number of engineering/geological surveys have been performed at the Chornobyl NPP site from 1967 through the 1990s. Before construction of the power plant, the absolute elevation of the bare ground varied from 114–116 m. The graded elevation for the surface of the construction site was 114 m, this elevation was adopted as relative elevation 0.00.

Before the Chornobyl NPP accident, the groundwater level was approximately 9.6 m below grade—absolute elevation 104.4 m. Over the years since the accident, the groundwater has risen, and is now at approximately 110.0 m.

The overall hydrogeological environment in the vicinity of Chornobyl NPP is determined by the water level in the cooling pond, which is virtually stable at 110.5 m, as well as cyclical changes in groundwater level. These cyclical changes are due to natural changes in groundwater infiltration and reach approximate maximum values of 1–3 m, depending on the relationships between the various cycles, which have periods of 3, 11, 33, 90 years, or even longer.

### 3.2.1.4 Impact of Technological Factors

- *Structural collapse of Building B:* The probability of Building B structurally collapsing (and, in particular, vent stack collapse) is once every 10,000 to every 100 years.

The collapse of Building B does not automatically imply collapse of the Shelter. As reliable calculations are not available at this time, we shall assume that the probability of Shelter collapse for an initiating event which leads to collapse of Building B is 0.25.[5]

- *Effects due to operation of vibration-generating equipment in Unit 3:* The dynamic effects due to the Unit 3 turbine generators and main reactor coolant pumps do not have any substantial impact on the strength or stability of the main load-bearing structures in the Shelter. The dynamic loads due to operation of vibration-generating equipment in Unit 3 are at most 1.5–2% of the static load.

- *Aircraft impact:* All calculations involving aircraft impact on nuclear power facilities are required to use a 20-metric-ton aircraft traveling at 200 m/s. Such an event is estimated to have a probability of once every 1,000,000 years. Government regulations place restrictions on aircraft operations in the vicinity of NPP sites. Low-altitude aircraft operation (less than 3,300 m) within 3 km of the NPP site is prohibited without special permits.

### 3.2.1.5  Development of Degradation Processes

- *Corrosion of metal structures:* Corrosion processes lead to degradation of the strength of the metal structures that make up the roof, protective walls, and partitions. Nominal corrosion rates of 40–50 μm/yr were assumed for structural steel members used to build the Shelter. This corrosion rate leads to an estimated longevity of 80 years for the new metal structures.

- *Aging of concrete:* The concrete is cyclically exposed to alternating moisture and thaw cycles, which leads to corrosion damage, and, hence, a gradual reduction in the strength of the concrete. Concrete in the reactor building was exposed to the explosion and subsequent high temperatures following the accident. The effect of these temperature extremes has not been adequately quantified to permit a realistic assessment of the risk posed by failure of the existing cement structural members.

### 3.2.2  Monitoring Existing and Newly Constructed Shelter Elements

The Shelter monitoring system still does not include an official system for monitoring structural elements. Until quite recently, information on the con-

---

[5]Editor's Note: While this is a strongly held expert opinion, it is clearly unrealistic as it predicts a 100% probability of the Shelter's collapse in only 4 years!

dition of structural elements was generally obtained from visual inspection and surveying results to determine settling and tilt using special benchmarks installed on the walls and roof of the Shelter. No systematic predictive measurements have been made. The Shelter structural-element status monitoring system (developed by All-Union Scientific Research and Design Institute for Construction Management and the Shelter seismoacoustic monitoring system (developed by the Institute for Geophysics, Ukrainian Academy of Sciences, Kiev) have seen virtually no use for a variety of reasons.

In the past few years (1994–1996), researchers have conducted systematic inspections and functional evaluations of the Shelter hardware. In addition, new operating instructions and documents defining the parameters to be monitored and the monitoring sites to be selected have been written. Further, researchers have begun developing an integrated system for monitoring and predicting the status of Shelter structural elements.

### 3.2.3 Reliability of Major Load-Bearing Elements

Estimating the reliability of the multistory reinforced-concrete portion of the structure used to support the metal roof structure was quite difficult in the absence of important quantitative data.[6] Using normal weather conditions available in 1995, relative estimates for the reliability of the structural elements making up the outer Shelter shell show the foundation and substructure require no immediate measures to strengthen them for short-term use. The metal structures in the roof are generally sufficiently reliable. However, field studies indicate several localized defects that should be eliminated.

The relative estimates of the reliability (in the form of the index $\alpha$, the probability of failure, 1/yr) were calculated for the following fragments of the structure:

| | |
|---|---|
| Local support area for the Octopus girder | 0.0002 |
| Deaerator mezzanine framework | 0.000001 |
| West support under "Mammoth" girder | 0.0011 |
| East support under "Mammoth" girder | 0.0016 |
| Girders B1 and B2 (support at coordinate 50) | 0.13 |
| Girders B1 and B2 (supported by vent stacks) | 0.10 |
| Buttressed wall 0.0005 "Hockey sticks," north | 0.0001 |
| "Hockey sticks," south | 0.001 |

---

[6]Editor's Note: Where these data are unavailable, expert opinion has been applied to permit a quasi-quantitative estimate.

These results were obtained using a specially developed "retrospective method" [Nemchinov et al. 1997] rather than the classical methods of reliability theory (which cannot be used under these conditions). Of course, these estimated reliabilities will not remain the same under extreme conditions such as seismic events or tornadoes.

Using the retrospective method results, researchers generated a list of eight safety-related structural elements for the Shelter.

## 3.3 FISSILE AND RADIOACTIVE MATERIALS IN THE SHELTER AND AT THE SHELTER SITE

### 3.3.1 Pre-Accident Nuclear Fuel in the Reactor

Before the accident, Unit 4 contained nuclear fuel in the form of fuel assemblies in the reactor core, in the spent-fuel-assembly pool, at the fuel channel preparation location in the central hall, and at the fresh fuel preparation location. The total nuclear fuel content at these locations may have been 231.4 metric tons. The pre-accident nuclear fuel distribution is presented in Table 3.3.1 [Bogatov et al. 1994].

**TABLE 3.3.1.** Location and Characteristics of Nuclear Fuel in Unit 4 Before the Accident

| Location | Location Function | Uranium Content Nuclear Fuel, T | Burnup, MW x d/kgU |
|----------|-------------------|--------------------------------|--------------------|
| 504/2 | Reactor vault | 190.2 | 11 |
| 505/4 | Spent fuel assembly storage pool | 19.4[a] | 11[a] |
| 914/2 | Central hall | 2.3[a] | 0 |
| 503/2[b] | Fresh fuel preparation location | 19.5[a] | 0 |
| Total | | 231.4[a] | |

(a) Upper limit for this quantity.
(b) Nuclear fuel was removed from the preparation location (location 503/2) to the Chornobyl NPP fresh fuel storage facility in 1987.

### 3.3.2 Description of Pre-Accident Radionuclides Built Up in the Reactor Core

Beginning in 1986, a wide range of studies were conducted to calculate the quantity and composition of radioactivity accumulated in the pre-accident Chornobyl Unit 4 reactor core [see Bogatov et al. 1994; Bogatov et al. 1995; Kirchner and Noack 1988; Borovoi 1989; Begichev et al. 1990]. The studies show the total fueled core mass was 190.2 tons. Mean burnup was approximately 11 MW x d/kgU. The buildup calculations are based on the following source data:

- At the time of the accident, the reactor core contained 1,659 fuel channels with fuel assemblies, a single additional absorber, and a single unfueled channel. (Table 3.3.2 shows the distribution of fuel channels by burnup in an eight group approximation.)

- A significant number of the fuel channels were initially fueled channels with a burnup of 11–15 MW x d/kgU.

- A certain quantity of fresh fuel was also located in the core.

- Fuel-channel uranium mass was 0.1147 tons.

The specific activities of the fission and activation products were calculated using the WIMS [Askew et al. 1966] and CACH2 [Tataurov 1987] codes.

Following a calculation of the dependencies of the specific activities of produced radionuclides on burnup, the total buildup of radionuclides of interest within the reactor core was calculated. The buildup of the entire

**TABLE 3.3.2.** Distribution of Fuel Channels with Respect to Mean Burnup

| Group | No. of Fuel Assemblies | Burnup (MW x d/kgU) |
|---|---|---|
| 1 | 721 | 13.7 |
| 2 | 392 | 12.3 |
| 3 | 154 | 10.3 |
| 4 | 101 | 8.8 |
| 5 | 35 | 7.0 |
| 6 | 43 | 5.4 |
| 7 | 41 | 3.4 |
| 8 | 172 | 1.2 |
| Total no. of fuel assemblies | 1659 | |
| Average burnup | | 10.9 |

spectrum of isotopes and, particularly, the transuranic isotopes, has a highly nonlinear relationship to the burnup. Data containing the actual burnup values for each of the 1,659 fuel channels in accordance with a fueling diagram was input into a personal computer for this calculation. A custom program utilizing the relations generated by the WIMS and CACH2 codes calculated the mass of the radionuclides for each channel and then summed the mass over the entire reactor core. In addition, the program was capable of producing this information for any channel or group of channels. The next step in refining the calculations was to account for the power peaking factor in the RBMK-1000. A cosinusoidal nonuniform burnup was assumed in *Determination of the Nuclear Physics Properties of ChNPP Unit No. 4 Fuel* [Borovoy et al. 1991a]. A peaking factor of 1.4 was assumed; this factor is the ratio of maximum burnup to average burnup and takes into account all reactor (and fuel channel) vertical and radial nonuniformities.

### 3.3.3 Nuclear Fuel and Other Radioactive Materials Remaining in the Shelter (Integral Estimate)

By fall 1986, the measurements conducted by a number of scientific research institutes, state hydrometeorology services, and military organizations made it possible to estimate the fuel component release [Bogatov et al. 1995]. The fuel component release was $3 \pm 1.5\%$ of the total pre-accident core fuel [Kazakov and Berchii 1994]. These figures for the fuel component fallout (deposits) were subsequently confirmed between 1987 and 1989 by the analysis of thousands of soil samples: $3.5 \pm 0.5\%$.

However, the results for the volatile components were quite different. The $^{137}Cs$ release was estimated to be 12–30% of the total accumulated activity (which was $2.6 \times 10^{17}$ Bq), while the $^{131}I$ release was 20–50% (in 2.5) of the total accumulated activity [IAEA 1996b]. Figures 3.3.1 and 3.3.2 show, respectively, the activity of the fuel from Unit 4 and the residual heat released in the Shelter as a function of time.

### 3.3.4 Quantitative Estimate of Nuclear Fuel Modifications Inside the Shelter [Bogatov et al. 1995]

Studies conducted between 1986 and 1995 demonstrate that the irradiated nuclear fuel in the Shelter exists in four modified forms [Borovoi et al. 1994]:

- Core fragments
- Fuel particles or radioactive dust
- Solidified lava-like fuel-containing material
- Dissolved forms of certain radionuclides.

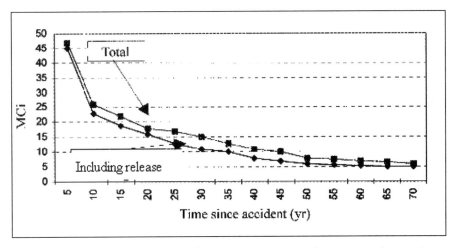

**FIGURE 3.3.1.** Activity of fuel in Shelter, as a function of time since the accident

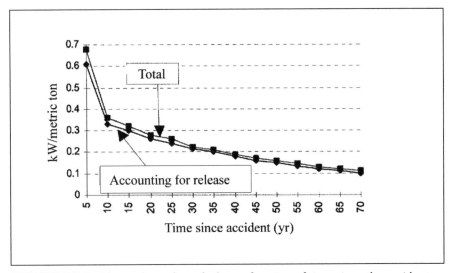

**FIGURE 3.3.2.** Heat release from fuel, as a function of time since the accident

Fuel distribution data are presented in Table 3.3.3 [Bogatov et al. 1995]. However, currently available information cannot be used to complete a quantitative mass balance for the core. This will require additional research. Only an upper and lower boundary on this value can be determined.

The majority of the core fragments were ejected during the explosion into the upper stages of the unit, specifically, into the central hall. The finely dispersed fuel (dust) is made of hot fuel particles that vary in size from

**TABLE 3.3.3.** Distribution of Fuel-Containing Material by Shelter Room/Location

| Room/Location (elevation) | FCM Classification and State | Estimate of Fuel Content of FMC (in metric tons of uranium) |
|---|---|---|
| Central hall (35.50) other upper unit locations | Core fragments (the majority are saturated with materials ejected during the active stage of the accident; lava-like FCMs may be located underneath the material) In the vicinity of System E - 10 - 36 | 60–70 |
| Southern spent fuel pool (18.00 - 35.50) | Spent fuel assemblies | ~20 |
| Majority of locations | Fuel dust. Hot fuel particles | ~10 |
| Subreactor locations 305/2 9.00) + 307/2 + System OR + reactor vault | Lava-like FCMs, core fragments | $75^{+25/-35}$ |
| Steam distribution corridor (6.00) allowing for FCM in the valves | Lava-like FCMs | $25^{+/-11}$ |
| Level 2 pressure suppression pool (PSP-2) | Lava-like FCMs | $8^{+/-3}$ |
| Level 1 pressure suppression pool (PSP-1) | Lava-like FCMs | $1.5^{+/-0.7}$ |
| 304/3, 304/3, 304/3, 304/3, "Elephant's Foot", etc. | Lava-like FCMs | $11^{+/-5}$ |
| Reactor unit, reactor building auxiliary system, and turbine hall locations | Water with dissolved uranium salts | ~3,000 m$^3$ water < 3 kg uranium |
| Fuel under cascade wall | Core fragments | Not identified |
| Fuel at site under concrete and crushed rock layer | Core fragments, fuel dust | $0.6^{+0.3/-0.2}$ |

FCM = fuel-containing material.

fractions of a micron to hundreds of microns. The fuel dust can be observed in virtually all locations throughout the facility. Solidified lava-like fuel-containing material was formed during the active stage of the accident (26 April–6 May 1986) from the high-temperature interaction between the fuel and structural materials of the unit. The lava-like material spread into the subreactor locations. In 1990, scientists discovered that water located in a number of the lower areas of the facility contained dissolved forms of uranium, plutonium and americium. These soluble compounds were caused by the breakdown of a variety of compounds created from the uranium dioxide fuel under multiple factors, the primary factor being the water that penetrated the Shelter.

3.3.4.1 **Core Fragments.** Significant quantities of isolated core fragments and accumulated fragments were observed around the destroyed Unit 4 reactor immediately following the accident. Fuel assemblies, fuel elements, and isolated parts of these components were ejected by the explosion onto the vent stack, the roofs of the deaerator mezzanine, the turbine hall, and the reactor building auxiliary systems as well as the roof of Unit 3 structures and in the central hall [Usaty et al. 1991; Dushin et al. 1992]. Some of the fragments were remediated following the accident. However, the majority of the fragments were not remediated. Researchers have performed several studies to determine the location of the fragments.

Several methods were used during accident remediation to deal with core fragments. Some of the core fragments from the site were moved to the debris pile and were buried in the cascade wall. Other core fragments were placed into containers with highly radioactive waste or were buried under the concrete and crushed rock layer surrounding the building. Some of the core fragments located on the roofs of buildings and the vent stack were discarded into the collapsed area of the building. Finally, a small quantity of core fragments remained on the vent stack areas and on the roofs; this was confirmed by the results of the 1990–1991 surveys.

The majority of the core fragments are probably in the reactor central hall. However, the material dropped by helicopter to put out the reactor core fire makes it impossible to directly observe significant accumulations of core fragments. Nonetheless, indirect facts such as the nature of the explosion, the large number of fragments ejected onto the roofs and around the reactor building, and the likely trajectories of the core fragments are consistent with this hypothesis. It is also confirmed by the rather significant number of fragments observed in the northwest and western sections of the central hall.

Researchers have conducted several studies to determine the accumulation of core fragments within the central hall. Initial studies were performed by specialists from the Kurchatov Institute [Usaty et al. 1991]. Further studies performed by specialists from the Kurchatov Institute in 1996–1997 used a highly collimated gamma detector, laser range finder, and video camera.

In 1992, a group from the Radium Institute, in conjunction with officials from the operational division of the Institute of Atomic Energy and the Chornobyl NPP, conducted a range of studies to determine the location of the primary gamma-emitters in the western section of the central hall. This group studied conditions in the vicinity of the displaced biological shield

(System E) and estimated the accumulation of core fragments to be between 10 and 36 metric tons. [Dushin et al. 1992].

A comparison of the 1992 gamma-scanning data to the video and later photo survey data suggests that the majority of the fuel is located in fuel channel remnants on the surface of System E. Approximately 1 metric ton of fuel is located on the walls of the central hall and other structural elements outside System E [Dushin et al. 1992].

Efforts have also been made to assess the location and condition of nuclear fuel in the spent fuel pool and new fuel storage areas. Periscopic observations and video surveillance in 1992 revealed that the panels containing fuel assemblies in the southern spent fuel pool continued to be suspended in their expected locations after the accident without noticeable damage.

Periscopic and video surveys revealed core fragments in the reactor vault. These core fragments were located on System OR. The number of such fragments is difficult to determine because of the collapse of the graphite masonry and concrete that was added to the shaft following the accident.

Approximately 10 accumulations of core fragments were discovered through visual surveys in the subreactor area. It is noteworthy that intact fuel pellets (uranium dioxide) from this location were identified from samples of the lava-like fuel-containing material indicating that there was a very broad range of temperatures in the area following the accident.

As the core fragments represent the most hazardous form of nuclear material from an occupational and environmental safety viewpoint, a conservative estimate, i.e., the upper limit on the quantity of this fuel modification, should be used. The total pre-accident uranium mass in the core, the southern spent fuel pool, and the central hall is estimated at 212 metric tons (this is a maximum value, neglecting any fuel assemblies removed from the fuel preparation area), while the uranium mass of the lava-like fuel-containing materials and fuel particles is estimated at 80 metric tons (the minimum value). The difference between these values (132 metric tons) after subtracting the ejected fuel (6 tons) is equal to 126 metric tons. This value represents an estimated upper limit on the uranium mass in the form of core fragments.

**3.3.4.2 Fuel Particles or Radioactive Dust.** A fairly large number of studies have been devoted to analyzing the physical and chemical composition of the radioactive dust (also called fuel particles or finely dispersed fuel) in various locations throughout the Shelter (see, for example, Khlopin Radium Institute 1989, 1991, 1992). Thorough measurements have not yet been performed; however, several estimates have been calculated for various locations.

Calculating the total amount of finely dispersed fuel on the surface of the central hall debris and other open surfaces under the roof of the Shelter provides a conservative estimate, resulting in an order-of-magnitude value of 1 metric ton (as of 1991). Other efforts to calculate the amount of radioactive dust in other Shelter locations have been based on measurements of the dose exposure rates in these locations. These calculations assumed that in those areas lacking noticeable quantities of core fragments or "fuel lava," the dose exposure rate was determined by the uniform layer of active dust remaining on the walls, floors, and ceiling of the room location. According to recent estimates, the Shelter contains approximately 34 metric tons of fuel in the form of radioactive dust (official report to SIP meeting on 20 January 1998 in Slavutych). The estimates on the amount of dust differ because different areas were analyzed. For the smaller estimate, only the open surfaces were analyzed. For the larger estimate, all of the surfaces were analyzed.

3.3.4.3 **Soluble Fuel Forms.** For some time, it was believed that none of the compounds created from the uranium dioxide fuel during the accident were soluble. However, in September 1990, bright yellow spots and pools were discovered on the surface of solidified lava flows in the steam separation corridor. Analysis of these spots revealed the presence of soluble uranium compounds. X-ray phase and microstructure analysis revealed the following compounds [Bogatov et al. 1990]:

- $UO_2CO_3$            Rutherfordine
- $UO_3 \cdot 2H_2O$         Eliantinite (Western name seems to be schoepite)
- $UO_3 \cdot 16CO_2 \cdot 1.19H_2O$    Studitite (Western name not identified)
- $UO_4 \cdot 4H_2O$
- $Na(UO_4)_2(CO_3)_3$

The plutonium isotope content in these new formations is hundreds of times lower while the mass concentration of uranium is several times greater than in the lava-like fuel-containing materials.

3.3.4.4 **Lava-Like Fuel-Containing Materials.** The high temperatures associated with the accident melted the zirconium fuel cladding and led to an interaction between the molten zirconium and the uranium dioxide, resulting in a uranium-zirconium-oxygen phase. When this phase interacted with structural materials (serpentine, concrete, and sand) as well as air, lava-like fuel-containing materials were formed.

Researchers working at Chornobyl NPP Unit 4 encountered this lava-like fuel-containing material for the first time in the fall of 1986. Subreactor location 217/2 was found to contain a large solidified mass, approximately 1 m wide that came to be called the Elephant's Foot (Figure 3.3.3). Analysis of the Elephant's Foot revealed that it consists primarily of silicon dioxide with other compounds as impurities, including uranium compounds. The mixture of radionuclides found in samples of the Elephant's Foot match those found in the irradiated nuclear fuel with an average burnup for Unit 4 [Anderson et al. 1993; Borovoi et al. 1990a; Borovoy et al. 1991b].

Lava-like fuel-containing materials were subsequently discovered in many subreactor room locations. This material contains significant amounts of pre-accident uranium from the reactor core and a significant number of reactor radionuclides.

**FIGURE 3.3.3.** Elephant's Foot, a lava-like fuel-containing material formation

***Formation and dissemination of the lava-like fuel-containing materials:*** The heat in the reactor core during the accident and afterwards may have melted portions of the uranium dioxide fuel and other materials in the core. The precise sequence of events that created the lava-like material is unknown. However, researchers do know the lava remained heterogeneous—much of the structural metal retained its composition while the ceramic mass interacted with structural materials as it flowed.

As the mass of the melt increased, it spread along the floor of location 305/2, reached the edges of the steam relief valves, migrated downward through the valves, and passed into the upper and lower portions of the

Suppression Chamber, part of the confinement systems designed to isolate steam in a design-basis accident. These locations include the steam distribution corridor and two levels of the pressure suppression chamber, which are located at elevations 6.00, 3.00 and 0.00, respectively. The melt could also flow horizontally, since a breach (or a burn-through hole—this is not known precisely) formed in the wall between locations 305/2 and 304/3.

***Physicochemical properties and radionuclide composition of the lava-like fuel-containing materials:*** *In* studying the lava-like materials, several materials are easily identified. The most common materials can be arbitrarily categorized as follows:

- Black ceramic with 4–5% uranium concentration by mass

- Black ceramic with 7–8% uranium concentration by mass

- Brown ceramic with 9–10% uranium concentration by mass

- Slag and pumice-like fuel-containing materials.

Figure 3.3.4 shows the percentage uranium for the lava-like fuel-containing material (based on data from 225 samples), while Figure 3.3.5 shows the fuel burnup for the lava-like materials.

***Degradation of lava-like fuel-containing materials:*** Degradation of the lava-like fuel-containing masses is significant because it represents a transition from a stable to a less stable form of radioactive material, increasing the amount of activity available for release to the environment. The degradation begins with a loss of integrity in the surface layers of the lava-like fuel-containing materials. This can accelerate processes that depend on the fuel-containing material contact area and the interacting medium: dissolution, erosion, etc. Thus, analyses were conducted on the lava-like material decomposition. The analyses have revealed three decomposition processes:

- Surface decomposition of the lava-like materials and the formation of radioactive dust on these materials

- Leaching of radionuclides from interaction between the "lavas" and water, and the formation of new chemical compounds on their surfaces

- Matrix cracking due to interior stresses.

An initial attempt to take a sample of lava-like fuel-containing material from the Elephant's Foot demonstrated its significant strength; small weapons were required to separate a sample from the surface of the formation in 1987. Repeated sampling from the Elephant Foot, beginning in 1990,

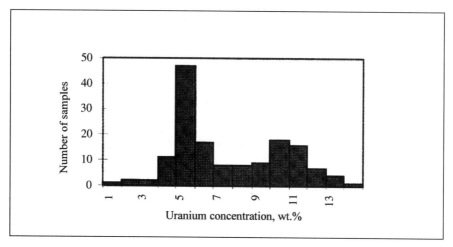

**FIGURE 3.3.4.** Fuel content (as U) in pure lava

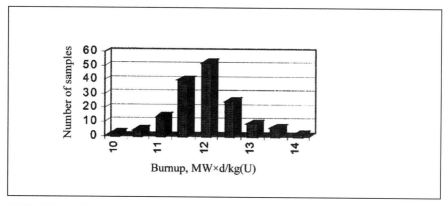

**FIGURE 3.3.5.** Fuel burnup in lava-like fuel-containing samples

no longer requires significant effort. Attempts to remove surface contamination from the Elephant's Foot, by means of a glue-treated wad, easily separated an upper 1–2 cm layer. At present, the lava degradation processes have caused the Elephant's Foot to nearly lose its initial shape, to slide down and settle. According to observers, the lava-like fuel-containing materials at room locations 304/3, 305/2, etc. have undergone similar changes.

Processes leading to lava-like fuel-containing material embrittlement have not been the focus of a special study. The most likely mechanism is associated with cooling of the lava-like materials and their impregnation with water: observations have revealed that the temperature of even small lava-like materials differs only slightly from the ambient air temperature. In winter, freezing water in the pores and microcracks of the lava-like materials may lead to their cracking. Moreover, microscopic analyses have revealed the severe inhomogeneity of the lava-like materials on the microlevel. Interior stresses caused by differences in the coefficients of thermal expansion of the components may arise as the temperature of the lava-like materials decreases substantially below the fuel-containing material solidification temperature. Because of the heterogeneous quality of the lava-like fuel-containing material, the coefficient of thermal expansion and even melting temperatures can vary significantly. Given this situation and the uncontrolled rate at which the lava-like fuel-containing material cooled, there are significant residual natural stresses that can accelerate the degradation of the material.

## 3.4  NUCLEAR SAFETY ASSESSMENT OF THE SHELTER

This section examines general issues relating to nuclear safety of the Shelter and analyzes critical parameters of the fuel-containing material clusters.

### 3.4.1  General Concepts of Safety Assessment

During the last 10 years, the fuel-containing material in the Shelter was exposed to a broad range of external environmental influences: a large quantity of water has penetrated the facility; major construction operations that could have been accompanied by structural shocks to the building were implemented (1987–1989); and, in the summer of 1990, the Shelter experienced a magnitude 4 earthquake. The observations and studies conducted during the 10 years following Shelter construction have not revealed any increase in the criticality of the fuel-containing material accumulations. (The detection of transient increases in neutron count rate on detectors placed at these locations occurred in 1990 at room location 304/3 and in 1996 in room location 305/2; these detections are considered "potentially suspicious.") The evidence from these observations and studies all suggests that given the current state, the fuel-containing material accumulations are subcritical.

We, the authors, believe a complete nuclear safety assessment of the Shelter including physical measurements to determine the margin of nuclear subcriticality should be conducted to determine if certain events could lead to a criticality within the facility; however, such an assessment cannot be fully implemented at present. The reason is the lack of a comprehensive data set on the Shelter; specifically, on the fuel-containing material in the facility. It is unlikely that such a physical assessment could even be performed given the complex geometry and hazardous radiological conditions in the vicinity of the lava-like fuel-containing material. Hence, an analysis can only be based on available information and expert determinations of unknown parameters. Following future research, estimates will more closely approximate reality, and it is possible that some of the countermeasures will turn out to be redundant.

### 3.4.2 Nuclear Safety Studies and Conclusions

The primary document describing nuclear safety at the Shelter is *Technical Substantiation of Shelter Facility Nuclear Safety*, issued by the Kurchatov Institute in 1990 [Belyaev et al. 1990]. However, research conducted since the safety study was necessary to substantially revise and partially reconsider the conclusions of the document.

3.4.2.1 **Technical Substantiation of Nuclear Safety.** The *Technical Substantiation of Shelter Facility Nuclear Safety* study summarizes experimental data and calculations obtained since the accident through the middle of 1990 concerning the fuel-containing material at the Shelter. All research has been conducted on the surface of the fuel clusters as extracting highly radioactive core samples is difficult because of limited access to the interior. Using this research, the study draws conclusions concerning the nuclear safety of the facility. A number of organizations in Russia, Ukraine, and Belarus were actively involved in this study. Two qualitative findings were stated in *Technical Substantiation of Nuclear Safety*. First, "a range of diagnostic measurements have indicated that all fuel-containing materials at the Shelter have been subcritical since the end of the active phase of the accident." Second, all fuel-containing material modifications are substantially subcritical if water is absent.

Quantitative characteristics of the neutron flux multiplication factor ($k_{eff}$ also known as k-effective) for the lava-like fuel-containing materials obtained from experiments are as follows:

- Less than 0.4 (passive methods using measurements of spontaneous fission only)

- Less than 0.7 (active methods using measurements of subcritical multiplication by applying an external neutron source).

In evaluating the possibility of criticality in the presence of water in the Shelter, researchers confirmed that all fuel-containing material modifications identified in any geometric combination of dimensions permitted by the Shelter locations are substantially subcritical in the absence of water.

Estimates of the criticality of mixtures of lava-like fuel-containing materials and water revealed that for the test lava compound, $k_{eff}$ is always much less than 1 (based on results obtained for surface samples). Two barriers were identified that prevented the penetration of water into the lava-like fuel-containing material clusters:

- A thermal barrier (estimates indicate that the larger lava-like fuel-containing materials would have to have an interior temperature significantly exceeding 100°C as the surface measured temperature was 60–70°C)

- A water-impermeable vitreous surface.

The test water samples taken from the Shelter revealed an additional safety barrier. The water contained neutron absorbers—boron and gadolinium salts, which dissolved into the water as it flowed over the material in the central hall (or as water was introduced during operations).

The final conclusion reads, "it can be determined that at present the Shelter is a nuclear-safe facility." However, analyses of the dynamics of fuel-containing material behavior revealed possible problematic trends that may lead to a decrease in the nuclear safety criticality margin and a review of the conclusions that the facility is safe.

### 3.4.2.2 Recent Studies on Shelter Nuclear Safety.

A substantial range of experimental analyses and calculations have been conducted over several years (1990–1997), including studies directly focused on the nuclear safety of the Shelter. These studies have provided significant results that made it necessary to substantially revise and partially reconsider the conclusions of the *Technical Substantiation of Nuclear Safety* [Belyaev et al. 1990].

In 1992–1993, analyses of the fuel-containing material samples taken from the subequipment room location 305/2 revealed unmelted reactor core fragments. Moreover, visual observations revealed core fragments in direct contact with the lava. Hence, the nuclear safety calculations and assessments then had to account for a new "lava + reactor core fragment + water" mixture, which in certain cases was potentially more reactive than "lava + water" only.

At the same time, a number of safety barriers preventing criticality (and predicted in the *Technical Substantiation of Nuclear Safety*) became noticeably lower. Calculations and experience indicated significant cooling of the lava, with cracking and conversion to a water permeable structure.

In recent years, as the decay heat from the nuclear fuel remnant within the Shelter decreased, the quantity of condensed water forming on cold surfaces within the Shelter rose noticeably. The water has not passed through the majority of materials and may not contain neutron-absorbing elements [Bogatov and Gavrilov 1996].

### 3.4.3 **Critical Parameters of Fuel-Containing Material Clusters**

Four fuel-containing material modifications were discussed in preceding sections: reactor core fragments, radioactive dust, lava-like fuel-containing material, and aqueous solutions of uranium salts. In principle, by accounting for the two moderators (graphite and water), it is possible to obtain at least 10 different mixtures. A more feasible approach is discussion of the critical parameters for each of the fundamentally possible fuel-containing mixtures.

3.4.3.1 **Reactor Core Fragments.** The risk that a self-sustained chain reaction would reoccur in the Unit 4 reactor core hung over researchers through mid-1988. This was because a comparatively small portion of the core (approximately 150 fuel channels of the original 1,600 fuel channels) containing graphite moderator could achieve criticality in the absence of absorber rods. Such a quantity of fuel had survived the accident although the geometry and moderation of this quantity remained unclear. In May 1988, by using boreholes drilled from decontaminated areas of the unit through concrete walls and other obstacles, researchers were able to observe the reactor vault. They established that no reactor vault or reactor core geometry as such existed [Borovoi et al. 1990b]; thus, a self-sustained chain reaction could not be initiated by a large core remnant.

With this question solved, researchers turned to the question of the hazard of reactor core fragments ejected from the explosion and the possibility that the fragments formed their own "cluster." While this question cannot be answered without some reservation, it can be stated that all clusters that have been accessible to observation to date pose no nuclear hazard.

Specialists from the Institute of Nuclear Energy calculated the critical parameters of a cylindrical configuration with a height equal to the half height of the RBMK fuel channel active region (342.8 cm), because as the fuel channels were ejected from the core the most likely location for damage was at the steel-zirconium junction and the structural link between two fuel assemblies. The problem was calculated for a "dry" system and a system filled with water for two burnups: 0 and 11.5 MW x d/kgU. In the criticality calculations for the case where the cluster is filled with water, the water was assumed to fill all interfuel element spaces as well as the gap between the fuel channel and the graphite.

The neutron multiplication factors in an infinite medium ($K_{inf}$) for such systems obviously exceed unity. The minimum number of fuel assemblies required to achieve a critical mass is of interest. Table 3.4.1 shows the relationship between $k_{eff}$ and size for new fuel and that which has been exposed to 11.5 MW x d/kgU.

Calculation results are also shown in Figure 3.4.1. This graph shows Keff plotted as a function of the radius of the fuel assembly. Fuel burnup is 11.5 MW x d/kgU.

In the absence of other information, it was entirely possible that a hazardous cylindrical cluster containing burned-up fuel approximately 5 m in diameter and 3.5 m in height could be located under a layer of materials in the central hall.

**TABLE 3.4.1.** Neutron Multiplication Factors in Reactor Core Fragments of Cylindrical Geometry with RBMK Design Structure

| No of. "Half "Height Channels" | $R_{cyl}$, cm | $K_{eff}$ (dry)/$K_{eff}$ (water) | |
| --- | --- | --- | --- |
| | | **Fresh fuel** | **11.5 MW x d/kgU** |
| 9 | 39.2 | 0.29/0.52 | 0.21/0.43 |
| 33 | 76.4 | 0.78/0.86 | 0.61/0.71 |
| 113 | 141.3 | 1.08/1.00 | 0.88/0.84 |
| 261 | 214.6 | 1.18/1.05 | 0.97/0.88 |

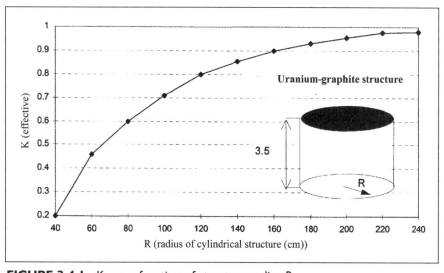

**FIGURE 3.4.1.** $K_{eff}$ as a function of structure radius $R$

Similar results were obtained in calculations by Kurchatov Institute specialists [Brodkin 1993]. A rectangular reactor structure comprised of $N \times N$ fuel assemblies at height $H$ under exposure to water was examined. The results are presented in Figure 3.4.2. The significant possible impact on the critical parameters of the clusters in the central hall resulting from even trace quantities of boron or gadolinium in the water or other materials has already been discussed. The calculations indicate that the existence of boron in the water increases the quantity of fuel required to achieve a critical mass by roughly a factor of 2. The eastern section of the central hall contained approximately 20 new fuel assemblies before the explosion on 26 April 1986. The explosion could have caused them to fall, be ruptured, and form a sufficiently compact system of fuel element fragments. If water penetrated into the system, it could by various means form a critical mass. The resulting estimated minimum diameter for a critical medium of fresh fuel and water turned out to be approximately 1 m.

The Institute for Radioenvironmental Problems in Belarus has estimated criticality for homogeneous combinations of fine fuel element fragments (fuel dust) and graphite dust. This mixture was analyzed for two fuel concentrations: 10% and 18% uranium dioxide (by mass). An average burnup of 11.5 MW x d/kgU was assumed. Produced fission products that are neutron absorbers were neglected in the spent fuel calculations to obtain maximum values of $k_{eff}$. A spherical model (as the most hazardous model)[7] was considered for this same reason. A concrete layer 0.4 m in thickness was considered as the reflector.

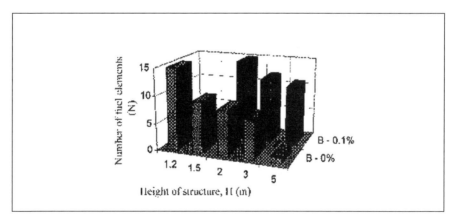

**FIGURE 3.4.2.** Critical parameters for an $N \times N \times N$ fuel-element structure with water alone and water containing 0.1% boron

---

[7]Editor's Note: As the surface-to-volume ratio for a sphere is lowest when compared to other geometries, this represents the worst-case scenario from a nuclear criticality safety standpoint.

Homogeneous spherical clusters containing burned-up fuel remain sub-critical or have dimensions that cannot be achieved within the observable fuel-containing material clusters. Any shape of the system other than spherical will result in a lower value of $k_{eff}$. In the case of a spherical mass, the minimum fuel and graphite masses (for fresh fuel) will be 1.1 and 30 metric tons, respectively, within a sphere 3 m in diameter; such a sphere cannot exist within any of the surveyed fuel-containing material locations.

3.4.3.2 **Fuel Dust.** Fuel dust is distributed over many locations within the Shelter. In those locations accessible to researchers, the dust appears as a thin layer and presents no nuclear hazard. However, it can be assumed that fuel dust has accumulated (or is accumulating) in locations that are inaccessible to observation. This may have occurred in upper locations (most likely the central hall) during ejection of fuel from the unit in the active phase of the accident. Dust may accumulate on the surface of the decomposing lava-like fuel-containing materials [Bogatov et al. 1991] or may be trapped in a lower location, having been transported to that location by water.

To estimate the quantity of dust (as uranium) that would need to have accumulated during the release to represent a realistic hazard in the future, the following must be considered:

- Fuel burnup in a dust accumulation would have to average 11 MW x d /kgU, as there is no burn-up separation mechanism in the release.

- A fuel dust accumulation in principle cannot be a nuclear hazard in the absence of a moderator. Water is a natural choice as a moderator because it could penetrate the fuel over time (as it cools).

- The fuel was ejected during a 10-day period, during which a variety of materials were dumped into the reactor wreckage. It can be assumed that the fuel dust exists in combination with other materials.

According to WIMS code calculations, the minimum diameter of a critical sphere, comprised of a mixture of sand, water, and uranium dioxide (50% by weight) and a burnup of 11 MW x d /kgU is 2.8 m. The critical mass of uranium dioxide in the sphere is 14 metric tons. This value substantially exceeds the integral estimates of the total fuel dust in the Shelter.

3.4.3.3 **Lava-Like Fuel-Containing Material.** The multiplicative properties of the uranium dioxide + silicon dioxide compositions were generally considered in the calculations [Bogatov and Gavrilov 1996]. These compositions were considered because they represent a "conservative" approximation of the actual lava-like fuel-containing materials, as neutron absorption is lower in silicon dioxide than in the actual set of elements (e.g., silicon, potassium, calcium, magnesium, aluminum, iron, and zirconium) encountered.

In the absence of a moderator, these urania mixtures and compounds are entirely safe from a nuclear criticality standpoint. Hence, the calculations were run for moist mixtures. The issue of what degree of penetration of water into the lava is realistic is discussed below.

As discussed in Section 3.4.2, one of the most important issues is whether or not a lava + water mixture actually forms. In the early years following the accident, water penetration into the lava clusters was inhibited by two barriers: a thermal barrier and the water-impermeable vitreous surface. Today, these safety barriers have been substantially reduced. The water that has penetrated the Shelter has led to severe destruction of the lava, and it can no longer be considered a "solid, vitreous" material.

3.4.3.4 **Aqueous Solutions of Uranium Salts.** Uranium appears in aqueous solutions gathering inside the Shelter as a whole range of complex salts. The solubility limit of these uranium salts may reach several grams per liter. The uranium isotope content as a function of sampling time and location varies over a broad range, up to tens of micrograms per liter.

The concentration of uranium is quite low in the aqueous solutions; this means that the aqueous solutions are very far from reaching hazardous concentrations; for an enriched (2%) uranium content in water of 10 g/L, $k_{eff}$ will only be 0.03. This value is even lower once the presence of boron is taken into account; boron appears in virtually all the water samples. The influence of boron was confirmed by calculations performed at the GRS Institute (Germany) for the uranium + water + absorber system [Gmal 1994].

### 3.4.4 **Nuclear Hazard Classification of Shelter Locations**

As noted previously, all of the research and long-term experience at the Shelter indicates that, given the situation as it now exists, the Shelter's fuel-containing materials are subcritical. An increase in fuel-containing material criticality and, ultimately, a self-sustained chain reaction can only occur if the fuel changes state. A fuel changes state when the fuel-containing material changes composition and new compositions are created for which the neutron multiplication factor will be greater than for the existing compositions. Such a change may be caused by a variety of initiating events.

Using the data from the preceding section, the fuel-containing material compositions currently at the Shelter are divided into four groups:

- Group 1: actual compositions and possible related compositions that cannot sustain a chain reaction because of nuclear physics, geometrical conditions, or other reasons

- Group 2: actual compositions, and those that may form on their basis, for which a self-sustained chain reaction is fundamentally possible, although it involves a confluence of such low-probability events that effectively no known nuclear hazard exists

- Group 3: new compositions that may form and in which a self-sustained chain reaction is fundamentally possible, although a number of known safety barriers exist to this outcome

- Group 4: new compositions that may form and in which a self-sustained chain reaction is fundamentally possible and for which there is no reliable information concerning the existence of safety barriers.

The locations at the Shelter where these or other compositions may develop are also divided into the same nuclear hazard groups.

Table 3.4.2. describes the primary locations in the Shelter that contain fissile material [Bogatov and Gavrilov 1996].

**TABLE 3.4.2.** Summary of Primary Shelter Locations Containing Nuclear-Hazardous Fissile Material

| No. | Location | Possible Nuclear-Hazard Composition | Formation Mechanism | Safety Barriers | Nuclear Hazard Group |
|---|---|---|---|---|---|
| 1. | 012/15 | None | None | a, b | 1 |
| 2. | 012/7 | None | None | a, b | 1 |
| 3. | 217/2 | None | None | a, b | 1 |
| 4. | 301/5 | None | None | a, b | 1 |
| 5. | 301/6 | None | None | a, b | 1 |
| 6. | 303/3 | None | None | a, b | 1 |
| 7. | 210/7 | None | None | a, b | 1 |
| 8. | 304/3 | Critical sphere exposed to water ($D_{cr} = 2.8$ m; $M_{uo} = 14$ metric tons) | Total change in geometric parameters accompanied by storm-precipitation and local change in fuel concentration | a, b, c | 2 |

*continued*

**TABLE 3.4.2** Summary of Primary Shelter Locations Containing Nuclear-Hazardous Fissile Material *(continued)*

| No. | Location | Possible Nuclear-Hazard Composition | Formation Mechanism | Safety Barriers | Nuclear Hazard Group |
|---|---|---|---|---|---|
| 9. | 210/6 | Critical sphere exposed to water ($D_{cr}$ = 2.8m; Muo =14 metric tons) | Total modification of geometrical parameters accompanied by storm precipitation and local change in fuel concentration | a, b, d | 2 |
| 9. | 210/6 | Critical sphere with an FCM and fuel element volume ratio of 4–8 exposed to water ($D_{cr}$ =1.6 - 2.9m) | Mixing of the FCM at locations 305/2 and 210/6 caused by a collapse of System E accompanied by storm-level precipitation and local changes in fuel concentration | c, d, e | 2 |
| 9. | 210/6 | Critical sphere with an FCM and fuel element volume ratio of 4–8 exposed to water ($D_{cr}$ =1.6 - 2.9m) | Mixing of reactor core fragments in the lava at location 305/2 and the lava at location 210/6 accompanied by storm-level precipitation | c, d, e | 2–3 |
| 10. | South spent fuel storage pool | Spherical fuel-water composition | Breakdown of channels and mixing of fuel assembly end components with low-level burnup, exposure to water | c, f, g | 2–3 |
| 11. | Location | Spherical fuel + water composition for fuel with average burnup ($D_{cr}$ = 3.2 m; $\gamma UO2/\gamma H2O$ = 30/70 volume %) | Total change in geometric parameters and exposure to water | a, c, d | 2–3 |
| 11. | Location 305/2 and the reactor vault (lower section) | Spherical fuel-water composition for low-(lower level burnup fuel ($D_{cr}$ =1.3m) | Exposure to water | c, d | 4 |

*continued*

**TABLE 3.4.2** Summary of Primary Shelter Locations Containing Nuclear-Hazardous Fissile Material *(continued)*

| No. | Location | Possible Nuclear-Hazard Composition | Formation Mechanism | Safety Barriers | Nuclear Hazard Group |
|---|---|---|---|---|---|
| 11. | Location 305/2 and the reactor vault (lower section) | Critical sphere with FCM and fuel element volume ration or 4–8, exposed to water ($D_{cr}$ =1.6m–2.9m) | Exposure to water | c, d | 4 |
| 12. | Location 305/2 and the reactor vault (lower section: the "debris pile" at OR system) | Spherical fuel-water composition uranium-graphite water system | Initiating event and formation mechanism are unknown | c, g | 4 |
| 13. | Central hall | Spherical cluster comprised of spent fuel assembly "ends" and water ($D_{cr}$ =1m) | Storm-level precipitation | c, g | 4 |
| 13. | Central hall | Spherical composition consisting of destroyed fuel assemblies containing fresh fuel and water ($D_{cr}$ = 0.7 m) | Storm-level precipitation | c, g<br><br>c, g | 4 |
| 13. | Central hall | Spherical $UO_2$ + $H_2O$ + C composition ($D_{cr}$ =1 - 2m) | Storm-level precipitation | c, g | 4 |

## 3.5 EVALUATION OF THE RISK OF RADIOLOGICAL CONSEQUENCES FROM A POTENTIAL SHELTER COLLAPSE

The most serious consequences involve the collapse of the Shelter roof. Such an event would cause a large mass of air to move upward, entraining in its wake a mixture of dust and fuel particles currently contained within the Shelter.

### 3.5.1 **Analysis of the Dust Plume Formation Mechanism in the Event of a Collapse of the Shelter Roof Structures**

The Kurchatov Institute and Russian Academy of Sciences, Nuclear Safety Institute studied the consequences of the Shelter roof collapsing. This section presents a proposed scenario for the roof collapse and formation of a radioactive dust cloud. This scenario is based on an understanding of the inherent weaknesses in the mechanical design of the Shelter. The former central hall has a planar surface area of approximately 220 m$^2$ at elevation 35.5, and an area of 260 m$^2$ at elevation 47 (near the remains of the west wall). The walls have a surface area of 3,100 m.$^2$

A roof collapse might begin with the collapse of the remains of the western interior walls at high elevations. This would cause the western ends of the roof girder framework and supporting girders for the side elements (the "hockey sticks") to fall from elevation 58.1 to elevation 48–49. Because of this motion, the eastern ends of the girder framework might fall off the supporting surfaces on the ventilation building (Block "V"), which would cause the central portion of the roof to fall on the debris mound at elevation 43–49 m. The eastern ends of the longer supporting girders might remain in contact with the vertical surfaces of the vent stacks. The metal connector installed over the east beams in the central hall might also remain in position. The side pieces of the roof (the "hockey sticks") would fall inward because of the loss of support. The buttressed stage of the stepped wall (i.e., the cascade wall) on the north side would presumably remain intact up to elevation 53.5. The falling masses have a maximum energy of approximately 45 MJ. The amount of radioactive dust located under the falling structures is estimated to be approximately 500 kg (in terms of fuel).

Several uncertainties arise when calculating the propagation of the dust plume beyond the Shelter in such a scenario. These uncertainties involve the height to which the dust plume rises, the composition of the dust, and the mass of dust released. The calculations used conservative estimates for all of these factors [Bogatov and Gavrilov 1996].

The impact of the falling roof structures will cause the surface of the debris mound to vibrate and suspend dust particles. The debris mounds were created by dropping sand and other materials onto the exposed reactor core in an attempt to stop the fire. The particles desorbed from the surface of the debris mound in this fashion will then be swept up in the rarefied turbulent wake behind the falling structures. The air movement over the surface of the debris mound as air is displaced from the central hall may provide an additional desorption mechanism [Bogatov and Gavrilov 1996].

The dust plume formed in the turbulent wake will rise due to air flow from lower levels within the Shelter and/or a temperature difference relative to the surrounding air. Estimates indicate that the resulting dust plume will reach the height of the remaining vertical walls within approximately 50 s; the air is expected to be heated by an additional 1 degree in the turbulent wake.

The maximum amplitude of the oscillatory motion in the surface of the debris mound will be approximately 1.4 m/s, which will throw the particles up high enough to escape the laminar boundary layer. The further motion of particles with a diameter of approximately 10 $\mu$m may be treated like the motion of a turbulent plume at typical air speeds of about 1 m/s. In this approach, it is assumed that all of the particles on the surface of the debris mound will be entrained in the plume. In actual fact, neglecting adhesion forces will grossly overstate the expected amount of dust raised.

The dust suppression compounds regularly sprayed on the surface of the central hall were taken into account by comparing the experimental defla-tion coefficients for particles from treated and untreated surfaces. These results led to the assumption that only 10% of the available contamination (approximately 50 kg of fuel) would be raised. The expected ratio of the radioactive component to the nonradioactive component in the dust plume is 1:100, i.e., the total mass of dust in the plume is expected to be on the order of 5 metric tons.

The height to which the plume would be raised was estimated for two mechanisms: one in which the plume "floats" on air flowing up from lower regions within the Shelter, and one based on semi-empirical models for the flow of a heated plume out of a tube. The latter model, which yields a value of $h_{max}$ equal to approximately 50–60 m, was used as a conservative esti-mate. Adding in the height of the debris mounds in the central hall leads to an effective plume height of approximately 100 m.

As the resulting height and diameter of the plume are comparable in size to the Shelter itself, the aerodynamic shadow on the lee side of the building was taken into account; estimates indicate that approximately 20% of the radioactivity will fall within this region. For the actual dimensions of the Shelter, the shadow will have an actual extent of approximately 240 m.

### 3.5.2 **Assessment of the Impact of a Roof Collapse Accident**

3.5.2.1 **Input Data.** Based on the studies by Kurchatov Institute and Russian Academy of Sciences, Nuclear Safety Institute, the dust plume will rise approximately 100 m above ground level (Section 3.5.1). The plume itself will contain about 40 kg of finely dispersed fuel as approximately 10 kg of fuel particles will be trapped in the aerodynamic shadow and settle within about 240 m of the Shelter. Table 3.5.1 presents data on the expected activities of the most significant radionuclides remaining in the Chornobyl NPP Unit 4 fuel.

**TABLE 3.5.1.** Activities of Various Radionuclides in the Plume (~40 kg of fuel from Chornobyl NPP Unit 4)[a]

| Nuclide | Activity, Bq |
|---------|--------------|
| $^{90}$Sr | $4.1 \times 10^{13}$ |
| $^{90}$Y | $4.1 \times 10^{13}$ |
| $^{137}$Cs | $5.2 \times 10^{13}$ |
| $^{137}$Pm | $2.0 \times 10^{13}$ |
| $^{238}$Pu | $2.2 \times 10^{11}$ |
| $^{239}$Pu | $1.8 \times 10^{11}$ |
| $^{240}$Pu | $2.5 \times 10^{11}$ |
| $^{241}$Am[b] | $4.8 \times 10^{11}$ |

(a) Impurity washout by precipitation was ignored for these calculations.
(b) The value quoted for $^{241}$Am is for a date about 10 years after the accident. $\beta$ decay of $^{241}$Pu will cause the activity of this radionuclide to increase (approximately 6% per year over the next few years). The change in the activities of all other nuclides may be neglected over the near term.

For purposes of calculation, the wind direction was assumed to be equal to the dominant wind direction at the Chornobyl NPP site during the entire warm period—from the northwest, with a mean value of 4.2 m/s.

3.5.2.2  **Results of the Calculation.** To calculate the impact of the dust plume containing 40 kg of finely dispersed fuel, a Gaussian model for the atmospheric diffusion was used, and the model was implemented using the VOYAGE code. Figure 3.5.1 shows the distribution of contamination per unit area on a scale of 1:200,000. The total activity is essentially split between the radionuclides $^{90}$Y:$^{90}$Sr:$^{137}$Cs:$^{147}$Pm approximately in the ratio 1:1:1:0.5; the density of $^{137}$Cs and $^{90}$Sr deposition at the edge of the 30-km zone will be approximately 148 kBq/km.[2]

The ejecta will be most hazardous several kilometers downwind from Unit 4. Tables 3.5.2 and 3.5.3 present the theoretical statistics for external radiation from the plume and for inhalation doses, respectively. (Calculations run in accordance with ICRP-30 recommendations [ICRP 1979].)[8]

---

[8]Editor's Note: The authors use an unusual combination of calculation methods in this section—they combine ICRP-2 (ICRP 1959) based methodology with selected portions of ICRP-30 (ICRP 1979) and SI units. While this may be considered by many to be inappropriate, it has been presented in its unaltered form. A more rigorous approach would not alter the basic conclusion of this boundary analyses.

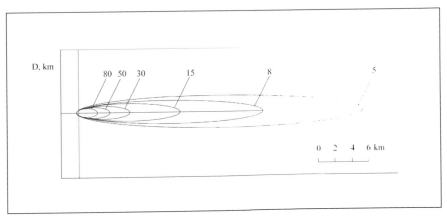

**FIGURE 3.5.1.** Contamination contours (Ci/km$^2$) for a hypothetical Shelter roof collapse (scale 1:200,000). *D* is the distance in km from the center of the hypothetical collapsed roof.

**TABLE 3.5.2.** Estimated External Dose Equivalent from Plume in the Event of a Shelter Roof Collapse

| Effective Dose Equivalent, Sv x 10$^{-6}$ | Area, km$^2$ |
|:---:|:---:|
| 1–2 | 0.8 |
| 2–3 | 0.4 |
| 3–4 | 0.4 |
| 4–5 | 0.4 |
| 5–6 | |
| 6–7 | 0.4 |

**TABLE 3.5.3.** Estimated Inhalation Dose Equivalent to Personnel in the Event of a Shelter Roof Collapse

| Effective Dose Equivalent, mSv | Area, km$^2$ |
|:---:|:---:|
| 5–10 | 0.8 |
| 10–15 | 0.4 |
| 15–20 | 0.4 |
| 20–25 | 0.4 |
| 25–30 | 0.4 |

Inhalation of radionuclides poses the greatest hazard. The annual effective equivalent dose is greater than 0.02 rems over an area of 0.8 $km^2$ within the plume. When the actual physical and chemical properties of the hot particles in the Shelter are taken into account, the results are likely to be much lower, and the results presented here should be treated as upper limits to the inhalation doses.

The aerodynamic shadow downwind from the Shelter, where approximately 10 kg of fuel dust acts as a bulk source, can be approximated by a rectangular parallelepiped with dimensions 240 m by 100 m by 60 m located downwind of the building.

Table 3.5.4 lists the expected atmospheric concentrations for the most hazardous (from the radiological point of view) transuranic elements in units of MPC. (The total value above the MPC may run as high as $4 \times 10^6$ during the settling/dispersal period.)

The characteristic longevity of the radioactive plume in the aerodynamic shadow of the Shelter is approximately 1 h. If personnel are present in this plume for 0.5 h, their inhalation exposure $n_a$ will be 900 times the annual limit.

Following deposition of the plume, radionuclides (mainly $^{137}Cs$) residing on the ground may produce a gamma-ray dose rate of 2.3 mSv/h at the center of the section under discussion.

**TABLE 3.5.4.** Expected Atmospheric Concentrations at the Chornobyl NPP in the Aerodynamic Shadow of the Shelter for the Most Hazardous (from the radiological point of view) Transuranic Elements (as of the end of 1996)

| Radionuclide | $^{238}Pu$ | $^{239}Pu$ | $^{240}Pu$ | $^{241}Am$ |
|---|---|---|---|---|
| Activity of plume, Bq/ L | 37 | 31 | 0.4 | 0.8 |
| MPC, Bq/L | $3.7 \times 10^{-5}$ | $3.3 \times 10^{-5}$ | $3.3 \times 10^{-5}$ | $1.1 \times 10^{-4}$ |
| Activity in units of MPC | $10^6$ | $9.4 \times 10^5$ | $1.3 \times 10^6$ | $7.4 \times 10^5$ |

### 3.5.3 GRS (German)[9] Analysis of a Roof Collapse Accident for Shelter

GRS has discussed a similar Shelter roof collapse scenario [Pretsh 1995]. The main differences relative to the model developed by Kurchatov Institute and Russian Academy of Sciences, Nuclear Safety Institute are as follows:

- Mass of the radioactive dust in the plume is a factor of 2 higher

[9]Editor's Note: GRS (Gesellschaft für Reaktorsicherheit) is Germany's central scientific-technical expert organization for all issues related to nuclear safety and nuclear waste management.

- Ratio of the radioactive mass of dust to the nonradioactive mass of dust is different
- Specific activity of the dust is slightly higher.

Slightly different constants were used in the atmospheric diffusion model. These results may be treated as an additional calculation that also included a full calculation of the organ-specific doses and the effective doses at large distances, including possible effects from rain-induced deposition of impurities. The radiation burden (organ-specific and effective dose) for a person within 2,000 m of the Shelter along the main axis of the plume was calculated using the GRS BEREG software.

The GRS study calculated uptake by inhalation during passage of the plume (for aerosol particles with activity mean aerodynamic diameter (AMAD) < 10 $\mu$m) and external gamma radiation from aerosols deposited on the ground; personnel exposure times of 1 day, 1 week, 1 month, and 50 years were used. The effects of the Shelter, plume "dilution," and other factors were taken into account. Particles with AMAD < 10 $\mu$m were ignored in all inhalation calculations.

The largest radiation burden[10] (both as the result of inhalation and as the result of external radiation from the ground) is received by the bone surface (maximum permissible dose [MPD] 300 mSv/yr), the liver (MPD 150 mSv/yr), the lungs (MPD 150 mSv/yr), and the red bone marrow (MPD 50 mSv/yr) (Figure 3.5.2) [Pretsh 1995]. The contribution of individual radionuclides to the radiation burden is tabulated in Tables 3.5.5 and 3.5.6.

**TABLE 3.5.5.** Percentage Contributions of Various Radionuclides to Radiation Burden via Inhalation (GRS Calculations)

| Nuclide | PercentageContribution, % |
|---------|---------------------------|
| $^{241}$Am | 32.8 |
| $^{240}$Pu | 24.2 |
| $^{238}$Pu | 18.5 |
| $^{239}$Pu | 17.4 |
| $^{90}$Sr | 6.7 |

[10]Editor's Note: The authors use the ICRP-2 (ICRP 1959) system of dose limitation in this section.

**TABLE 3.5.6.** Percentage Contributions of Various Radionuclides to Radiation Burden via Radiation from the Ground (GRS Calculations)

| Nuclide | Percentage Contribution, % |
|---------|---------------------------|
| $^{137}$Cs | 89.6 |
| $^{134}$Cs | 9.7 |
| $^{106}$Ru | 0.5 |
| $^{144}$Ce | 0.1 |

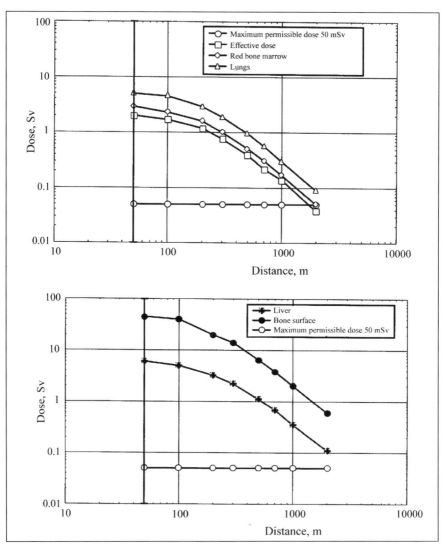

**FIGURE 3.5.2.** Inhalation doses (Category C weather, no rain, wind 2.53 m/h)

To perform the calculations, the GRS staff chose Pasquill weather category C, with no rain and a wind speed of 4.2 m at 100 m altitude (which corresponds to a wind speed of $u = 2.53$ m/s at the surface). A calculation was also performed for $u = 1.0$ m/s for the cases of zero precipitation and rainfall rates $l = 1$ mm/h and $l = 5$ mm/h. Additional calculations were performed for unstable weather (Pasquill Category A) and stable weather (Pasquill Category F), both with zero precipitation. See Table 3.5.7 for the Pasquill atmospheric stability classes.

For a wind speed of 2.53 m/s at 10 m altitude, effective inhalation doses of 2 Sv are observed 50 m from the source (Figure 3.5.2); at this dose, fatal effects may occur at a later date (unless individual protective gear is used). The effective dose only decreases to the MPD of 50 mSv at a distance of approximately 1,800 m. The radiation burdens are approximately four times higher for a wind speed of 1 m/s.

The largest increases in effective dose at small distances (from 50–100 m) occur for weather categories A and F. The most rapid decrease in dose with distance is expected for Category A, while the least rapid is expected for Category F. The doses for Category C with no rain and a wind speed of 1 m/s are only slightly higher than for those with rain.

Analyzing the dose values for a plume release in the different Pasquill weather categories, three trends can be seen. In calm stable weather, the airborne radionuclides are transported further and the concentration decreases at the plume "fans out." In all of the weather categories, the airborne particles settle to the ground (i.e., gravitational settling), with the larger particles

**TABLE 3.5.7.** Pasquill Atmospheric Stability Classes

| Wind Speed $u$ at Altitude 10m, m/s | Daytime for Various Insolation Levels | | | Nighttime | |
|---|---|---|---|---|---|
| | | | | Thin Clouds, Totally Overcast or Cloud Cover $\leq 4/8$ | Cloud Cloud Cover $\leq 3/8$ |
| | Strong | Moderate | Low | | |
| $u < 2$ | A | A - B | B | * | * |
| $2 \leq u < 3$ | A - B | B | C | E | F |
| $3 \leq u < 5$ | B | B - C | C | D | E |
| $5 \leq u < 6$ | C | C - D | D | D | D |
| $6 \leq u$ | C | D | D | D | D |

Note: A is Very Unstable; B is Moderately Unstable; C is Slightly Unstable; D is Neutral; E is Moderately Stable; and F is Very Stable.

*Official reference guides provided by some governments use Class F for wind speeds less than 2 m/s at nighttime.

Reference: Slade (1968)

settling more rapidly. This fallout increases the dose received from the ground, while decreasing the dose from inhalation. Finally, when it rains, more particles are deposited on the ground. This increases the dose from the ground, while decreasing the dose from inhalation. For example, if a heavy rain (a rate of 5 mm/h) is falling during the accident, the effective dose from the ground at a distance of 200 m will exceed the MPD within 12 hours. In dry weather, it takes 6 days to exceed the MPD.

### 3.5.4 Structural Element Collapse Scenarios for the Shelter Based on Calculations Performed by the Institute for Radioenvironmental Problems (Belorussian Academy of Sciences)

Several scientists from the Institute for Radiological and Environmental Problems (Belarussian Academy of Sciences) have studied the consequences of a roof collapse. They also considered gradual shifts in Systems L and E resulting from structural damage in the reactor compartment [Sharovarov et al. 1995].

The scenario considered by the Institute for Radioenvironmental Problems involves additional structural disturbances (possibly including movement of System E) not considered in either the Kurchatov Institute and Russian Academy of Sciences calculations or the GRS calculations, with a concomitant increase in dust generation. In the Institute for Radioenvironmental Problems scenario, collapse of the roof structures may be accompanied by collapse of the southeast wall of location 305/2 and downward motion of Systems L, D, and E. The lower edge of System E may even reach the top of the debris mound on the reactor base and transmit the pressure to the weakened floor of location 305/2, causing the latter to partially collapse into the underlying steam distribution corridor.

These researchers calculate the surface contamination resulting from propagation of the dust plume, assuming a moderate atmosphere and a surface wind speed of 2.5 m/s. A 2-metric-ton dust plume containing $9.6 \times 10^{13}$ Bq of radioactive material is raised to a height of approximately 100 m. The calculation was performed for two groups of particles: fuel particles with a mean median diameter of 30 $\mu$m and a density of 10.8 g/cm$^3$, and fuel-containing particles with a median diameter of 5 $\mu$m and a density of 6.0 g/cm$^3$.

Figure 3.5.3 shows the surface contamination contours. The total contamination at 50 km from the source is less than 74 kBq/m$^2$ (note that $^{90}$Sr and $^{137}$Cs each contribute about 30% of the radioactivity). Given the uncertainties and the simplifications made in this model, the results appear to be in fairly good agreement with the other calculations discussed.

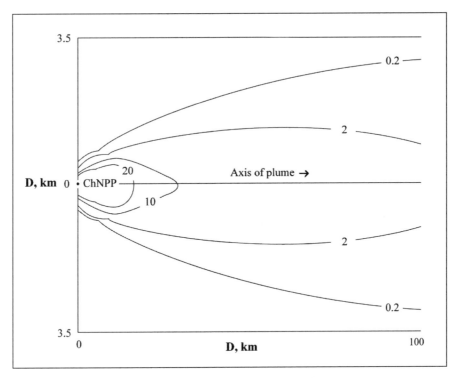

**FIGURE 3.5.3.** Surface contamination contours (Ci/km$^2$) (Institute for Radioenvironmental Problems calculation)

### 3.5.5 Conclusions on the Impact of a Roof Structure Collapse

The various studies of the consequences of a collapse of the Shelter roof are summarized in the previous sections. It should be noted that the calculations in these studies did not take into account the emergency dust suppression system, which expert estimates indicate should reduce the dust release by two orders of magnitude. Having reviewed these studies, we present the following conclusions regarding a potential collapse of the Shelter roof:

- At low wind velocities, the inhalation doses received during passage of the radioactive plume are extremely high (and exceed the MPD for personnel by two orders of magnitude) in the immediate vicinity of the source.

- Possible subsequent lethal outcomes for the several thousand personnel working at the Chornobyl NPP cannot be ruled out.

- The inhalation doses decrease below the MPD at distances greater than 10 km. Inhalation is not a hazard for the public outside the 30-km zone [Beskorovoinyi et al. 1997].

- The inhalation doses are largely due to long-lived transuranic elements, which means that the date of a potential Shelter collapse is not important to the calculation as long as it is within a few years of 1996.

- The radiation from the ground is largely due to long-lived $^{137}$Cs, which means that the date of the accident is not exceedingly important to the calculation in this case either.

- The doses due to radiation from ground contaminated during a potential collapse of the Shelter roof are an order of magnitude smaller than the inhalation doses (for an exposure time of 30 working days), and also decrease with distance from the Shelter.

- Rain has little effect on the doses; however, precipitation rate has a greater effect on the radiation dose from the ground than on the inhalation dose.

# 4
## ОТХОДЫ

# RADIOACTIVE WASTE
## *Storage and Disposal Sites*

V. Kukhar', V. Poyarkov, and V. Kholosha

After the accident, a large decontamination effort was performed around the four Chornobyl NPP units to rapidly isolate the large volume of contaminated soil, debris, wood, structural elements, demolished structures, equipment, and other contaminated materials by burial in mounds and trenches dug within the 30-km Exclusion Zone. These contaminated materials were buried to reduce radiation levels near the Chornobyl NPP (where reactors are still being used to generate power today) and to prevent atmospheric transport of radioactive dust and materials used to smother the fire (see Section 2.3 for accident chronology).

The materials were buried in two different types of sites: storage sites and disposal sites. Storage sites are considered temporary holding areas for the waste; the waste could be reclaimed or salvaged in the future. Disposal sites are considered permanent. The burial activities—especially the ones that occurred during the remediation efforts immediately after the accident—were not well documented. Reliable information concerning the volume and activity of radioactive waste being held at the waste sites was not collected when the sites were constructed. Site inventories began in 1990 and are continuing to date. The Science and Engineering Center for Integrated Radioactive Waste Management, which provided the materials on which this section is based, is performing the bulk of the inventory work [Ledenev et al. 1995].

## 4.1 DESCRIPTION OF RADIOACTIVE WASTE STORAGE AND DISPOSAL SITES

The bulk of the original waste burial work failed to comply with all of the requirements regarding engineered barriers against radionuclide migration, meaning that there is some likelihood that radionuclides will be transferred to the adjacent groundwater, and then be discharged into river basins.

The radioactive waste storage and disposal sites were established in 1986 and 1987 following completion of high-priority decontamination work in the 10-km zone around the reactor. The sites were combined into the following sectors on a territorial basis:

- Stroibaza Interim Radioactive Waste Storage Site, total area 1,250,000 m$^2$

- Ryzhii Les Interim Radioactive Waste Storage Site, total area 4,000,000 m$^2$

- Yanov Station Interim Radioactive Waste Storage Site, total area 1,280,000 m$^2$

- Pripyat Interim Radioactive Waste Storage Site, total area 700,000 m$^2$

- Neftebaza Interim Radioactive Waste Storage Site, total area 420,000 m$^2$

- Peschanoe Plato Interim Radioactive Waste Storage Site, total area 880,000 m2

- Kopachi Interim Radioactive Waste Storage Site, total area 1,250,000 m$^2$

- Chistogalovka Interim Radioactive Waste Storage Site, total area 60,000 m$^2$

- Buryakovka Radioactive Waste Disposal Site, total area 140,000 m$^2$

- Podlesnyi Radioactive Waste Disposal Site, total area 60,000 m$^2$

- Kompleksnyi Radioactive Waste Disposal Site, total area 20,000 m$^2$.

The Rozsokha Equipment Holding Facilities 1 and 2, a storage site for contaminated equipment, in the 30-km Exclusion Zone is described in Section 4.1.11 for the sake of completeness.

Expert estimates indicate that 6.3 x 10$^{15}$ Bq of radioactive waste from the Chornobyl NPP accident is currently being held at interim radioactive waste storage sites and 2 x 10$^{15}$ Bq is being held at radioactive waste disposal sites in the 10-km zone. The levels of surface contamination in the vicinity of the disposal sites are 10$^8$ Bq/m$^2$ for $^{137}$Cs, 7 x 10$^6$ Bq/m$^2$ for $^{90}$Sr, and 10$^5$ Bq/m$^2$ for $^{239,240}$Pu. The $^{90}$Sr content of the groundwater ranges from 0.18–3 Bq/L. The atmospheric aerosol activity adjacent to the ground ranges from 5.6 x 10$^{-3}$–9.6 x 10$^{-3}$ Bq/m$^3$.

The data in Table 4.1.1 imply a total of $1.78 \times 10^{15}$ Bq in the interim radioactive storage sites and $6.2 \times 10^{15}$ Bq in the radioactive disposal sites, for a grand total of $7.98 \times 10^{15}$ Bq.

**TABLE 4.1.1.** Inventories of Radioactive Waste Burial Sites in the 10-km Exclusion Zone (1996)

| Radioactive Waste Burial Site | Dose Rate in Monitored Zone | | Description of Waste | |
|---|---|---|---|---|
| | mR/h | $\mu$Gy/h | Quantity (volume), (m³) | Activity (Bq) |
| Stroibaza IRSS | 0.1–300 | 10–30,000 | $5.4 \times 10^5$ ($2.9 \times 10^5$) | $1.1 \times 10^{15}$ |
| Ryzhii Les IRSS | 0.3–100 | 30–10,000 | $2.5 \times 10^5$ ($5.0 \times 10^5$) | $4.8 \times 10^{14}$ |
| Yanov Station IRSS | 0.25–80 | 25–8,000 | $1.5 \times 10^4$ ($3.0 \times 10^4$) | $3.7 \times 10^{13}$ |
| Pripyat IRSS | 0.1–6.0 | 10–600 | $1.1 \times 10^4$ ($1.6 \times 10^4$) | $2.6 \times 10^{13}$ |
| Neftebaza IRSS | 0.2–5.0 | 20–500 | $2.2 \times 10^4$ ($2.2 \times 10^4$) | $6.2 \times 10^{13}$ |
| Peschanoe Plato IRSS | 0.2–6.0 | 20–600 | $1.8 \times 10^5$ ($1.0 \times 10^5$) | $3.7 \times 10^{13}$ |
| Kopachi IRSS | 0.1–3.0 | 10–300 | $0.9 \times 10^5$ ($1.1 \times 10^5$) | $3.3 \times 10^{13}$ |
| Chistogalovka IRSS | 0.02–5.0 | 2–500 | | $3.7 \times 10^{12}$ |
| Buryakovka RDS | 0.03–2.8 | 3–280 | | $2.4 \times 10^{15}$ |
| Podlesnyi RDS | 0.2–4.0 | 20–400 | | $2.5 \times 10^{15}$ |
| Kompleksnyi RDS | 0.07–6.0 | 7–600 | | $1.3 \times 10^{15}$ |

IRSS = interim radioactive storage site
RDS = radioactive disposal site

## 4.1.1 Stroibaza Interim Radioactive Waste Storage Site

This site includes a large fraction of the Chornobyl NPP commercial area (the plant's original industrial area) immediately to the east of the Yanov Station Interim Radioactive Waste Storage Site and extends to the Chornobyl NPP site itself, as well as the new construction support area. The Stroibaza Interim Radioactive Waste Storage Site includes approximately 350 to 400 localized facilities, repositories, and sites.

The top of the geological section where the Stroibaza Interim Radioactive Waste Storage Site resides contains nonuniform upper-Quaternary alluvial sand from the first fluvial terrace above the floodplain, with interbedded loam and loamy sand. This band has a maximum thickness of 30 m and is underlain by the Kiev marl formation. Groundwater is at a depth of 3–7 m.

The storage site is approximately 1,250,000 m² in area, and contains 284,000 m³/540,000 metric tons of waste. The local repositories are at a depth of 2.5–3.5 m, with a 0.3–0.5 m-thick covering of sand. This site contains metal, construction debris, wood, and soil. The site is estimated to contain a total of 1.1 x 10¹⁵ Bq.

The density of ¹³⁷Cs soil contamination varies from 1.8 x 10⁶– 3.7 x 10⁷ Bq/m². The maximum ⁹⁰Sr aqueous concentration from the monitoring wells is 60 Bq/L. The aerosol concentration in the atmospheric layer adjacent to the ground is 3.7 x 10⁻² Bq/m³.

### 4.1.2 **Ryzhii Les and Yanov Station Interim Radioactive Waste Storage Sites**

The Ryzhii Les and Yanov Station are adjacent to each other and are considered together because of their close proximity and the similarity of the waste contained in both sites. This site encompasses virtually the entire commercial area from the bank of the Pripyat River to Yanov Railway Station and contains almost exclusively contaminated trees.

The top of the geological section where these two sites are located consists of upper-Quaternary alluvial loamy and sandy deposits from the first and second terraces above the floodplain, with a total thickness of 30–45 m and nonuniform sand grains. The underlying layer consists of Kiev marl formation. The groundwater is at a depth of 1–5 m.

At Ryzhii Les, 60% of the contaminated materials were buried (with no water barriers) in trenches 3 m deep and the remainder was buried in mounds created from the local sandy soil. The Pripyat Dosimetric Monitoring Department has estimated the total activity in the trenches and mounds is approximately 4.8 x 10¹⁴ Bq, of which 2.6 x 10¹⁴ is ¹³⁷Cs, 2.2 x 10¹⁴ Bq is ⁹⁰Sr, and 5.18 x 10¹¹ Bq is ²³⁹,²⁴⁰Pu. The approximate mass of radioactive waste in the trenches and mounds is estimated to be 250,000 metric tons, with a volume of approximately 500,000 m³.

In 1991, All-Union Scientific Research, Design, and Experimental Design Institute for Power Technology (VNIPIET) studied mounds 1B through 6B and trenches 4T through 15T at Ryzhii Les (2% of the total storage site area). The gamma-ray dose rate in the burial site ranged from 40–140 mGy/L (40–140 mR/h). The specific activities of the radioactive waste were 1 x 10⁵–1.5 x 10⁵ Bq/kg for ¹³⁷Cs, 5.1 x 10⁴–3.7 x 10⁵ Bq/kg for ⁹⁰Sr, and 8.5 x 10²–1 x 10⁴ Bq/kg for the sum of all of the plutonium isotopes.

Groundwater and soil near the Ryzhii Les site is contaminated. The concentration of ⁹⁰Sr in the groundwater from monitoring wells near the trenches and mounds ranges from 0.3–3.2 Bq/L, while that in water from test pits located a few meters from the trenches may run as high as 3,000 Bq/L. A study done in 1991 by VNIPIET showed the groundwater activities in the mounds themselves were $3.7 \times 10^3$–$4.8 \times 10^4$ Bq/kg for $^{90}$Sr, 48 Bq/L–$1.2 \times 10^3$ Bq/L for $^{137}$Cs, and less than 48 Bq/L for plutonium isotopes and $^{241}$Am. The density of soil contamination is $1.8 \times 10^7$ Bq/m$^2$ for $^{90}$Sr, $4 \times 10^5$ Bq/m$^2$ for $^{239,240}$Pu, and ranges from $3.3 \times 10^7$–$5 \times 10^7$ Bq/m$^2$ for $^{137}$Cs.

At the Yanov Station site, the total activity stored is approximately $3.7 \times 10^{13}$ Bq. The depth of groundwater is approximately 2.5 m. Groundwater samples from monitoring wells have a $^{90}$Sr content of 0.25–1.2 Bq/L.

### 4.1.3 Pripyat Interim Radioactive Waste Storage Site

Adjacent to the city formerly occupied by Chornobyl NPP staff, this site has a total area of 700,000 m$^2$. Approximately 16,000 m$^3$ of scrap metal and household waste with a mass of about 11,000 metric tons is stored in this interim area. The total activity of the waste is approximately $2.6 \times 10^{13}$ Bq. The waste was buried at a depth of 1.5–2 m in sandy soil deposited as sediment, with a 0.3- to 0.5 m-thick covering of sand. The groundwater level is approximately 6 m.

To date, no information is available about the geology or possible environmental contamination at this site.

### 4.1.4 Neftebaza Interim Radioactive Waste Storage Site

Because this site is adjacent to the Pripyat River, it presents the highest danger of all of the sites for radionuclides entering the river and being carried downstream. Upon inspection of the site, 14 trenches containing 6,900 m$^3$ of radioactive waste were found to be continuously submerged and 25 trenches were found to be submerged during floods.

This 420,000-m$^2$ site contains approximately 22,000 m$^3$ of waste. The waste was buried in trenches dug 1.5–2 m deep. The trenches were dug in loamy sand and covered with a layer of loamy sand about 0.3–0.5 m deep. The groundwater level is approximately 6 m. The site contains approximately $6.2 \times 10^{13}$ Bq over the total area.

The density of surface contamination for $^{137}$Cs surrounding the site ranges from $5.6 \times 10^6$–$7.4 \times 10^6$ Bq/m$^2$.

### 4.1.5 Peschanoe Plato Interim Radioactive Waste Storage Site

This site, not included in records generated in 1986–1987, is located on a terrace built from sand dredged from the Pripyat Creek on the northwest bank of the creek and in the floodplain of the Pripyat River. Covering 150,000 m$^2$, this site contains 16 trenches (also called radioactive material burial sites). The trenches, dug in loamy soil, range from 15–660 m in length, 2–5 m in width, and 1–2 m in depth; the trenches are covered with a 0.3- to 0.5-m thick layer of local soil. The groundwater level is approximately 3 m.

Approximately 100,000 m$^3$ of radioactively contaminated soil is being stored in this area. The waste has a total mass of 180,000 metric tons and has a total activity of 3.7 x 10$^{13}$ Bq.

The $^{137}$Cs soil contamination within the aggradational terrace ranges from 1.8 x 10$^6$–7.4 x 10$^6$ Bq/m$^2$. The $^{90}$Sr concentrations in groundwater vary from 0.6–60 Bq/L.

### 4.1.6 Kopachi Interim Radioactive Waste Storage Site

This site is southwest of the Kompleksnyi Radioactive Waste Disposal Site, immediately adjacent to the Chornobyl NPP cooling pond. The burial area occupies approximately 1,250,000 m$^2$ and contains approximately 90,000 metric tons of radioactive waste, consisting mostly of metal structures and structural materials. The waste occupies approximately 106,000 m$^3$ and has a measured activity of 3.3 x 10$^{13}$ Bq. The materials are buried in loam to a depth of 2.5–3 m, with a 0.5–1 m covering of the same material. The groundwater level is 4–6 m.

To date, no information is available about the geology or possible environmental contamination at this site.

### 4.1.7 Chistogalovka Interim Radioactive Waste Storage Site

This site is in a former clay quarry on the northwest slope of the Chistogalovka Ridge near the Chistogalovka village. The upper geological section consists of nonuniform alluvial quartz sand belonging to the second terrace above the floodplain, with interbedded loamy sand and moraine till deposits. This band has a maximum thickness of 50 m and is underlain by the Kiev marl formation. The depth to the bottom of the former quarry varies from 6–8 m. Groundwater is at a depth of 3–15 m. No special water barriers were installed at the burial site.

This site was used to store household waste from the city of Pripyat, trees, soil, and construction waste. The radioactive waste has a measured activity of 3.7 x 10$^{12}$ Bq. The activity of $^{137}$Cs and $^{90}$Sr in the repository is generally assumed to be identical: 1.85 x 10$^{12}$ Bq. The soil in the vicinity of the site has a surface contamination density of 3.7–5.6 x 10$^6$ Bq/m$^2$. The maximum exposure due to the buried materials was 50 mGy/h (5 R/h).

### 4.1.8 **Buryakovka Radioactive Waste Disposal Site**

This site is located on Chistogalovka Ridge clay 6 km south of the village of Buryakovka. The upper portion of the geological section consists of middle-Quaternary fluvioglacial super-moraine deposits of fine-grained quartz sand. The deposit is approximately 50 m thick. The underlying, weakly permeable horizon consists of Kiev marl formation. The groundwater level is 20–24 m.

Medium-level radioactive waste (with maximum exposure rates of 10 mGy/h [1 R/h] at a distance of 1 m) from the Chornobyl NPP and population centers throughout the Exclusion Zone are disposed of at this site. The disposal site—with an area of 140,000 $m^2$ and a volume of 22,000 $m^3$—was constructed to a design generated by the VNIPIET. The repository currently contains approximately 11,000 metric tons of radioactive waste, including equipment, tools, and machinery contaminated during the accident remediation; metal; structural members; hardware; and waste produced by households in the area. For information on the economic losses from contaminated equipment buried at this site, see Section 8.1.2. A clay liner is used as a water barrier. As of 1 December 1996, the total radioactivity stored at the site was $9.7 \times 10^{14}$ Bq, and the on-site exposure rate varied from 0.4–14 µGy/h (0.04 –0.14 mR/h). Twenty-four trenches are currently filled.

The surface density of soil contamination in the vicinity of this disposal site is relatively low and varies from $7.4 \times 10^4$–$6 \times 10^5$ Bq/$m^2$ for $^{137}$Cs, 2.2 $\times 10^4$–$3.3 \times 10^4$ Bq/$m^2$ for $^{90}$Sr, and $2.2 \times 10^3$–$8.9 \times 10^3$ Bq/$m^2$ for $^{239,240}$Pu. The $^{90}$Sr content of groundwater from monitoring wells varies from 0.7–2.4 Bq/L. The maximum total activity concentration of aerosols in the atmospheric layer adjacent to the ground is 0.2 Bq/$m^3$.

### 4.1.9 **Podlesnyi Radioactive Waste Disposal Site**

This site is located on the right bank of the Pripyat River. The upper portion of the geological section consists of upper-Quaternary alluvial deposits (a 30-m-thick layer) above a 15-m-thick weakly permeable horizon consisting of Kiev formation marls and clays. The site is modular (based on a VNIPIET design); the site contains two modules, A-1 and B-1, with geometric dimensions of 55 $m^3$ by 26 $m^3$ by 8 $m^3$.

The total site is approximately 60,000 $m^2$ in area, with an approximate capacity of 10,000 $m^3$. As of 1990, Module A-1 contained 2,650 $m^3$ radioactive waste with specific activities of $2.2 \times 10^{11}$–$3 \times 10^{11}$ Bq/kg and Module B-2 contained 1,310 $m^3$. Approximately 2,500 $m^3$ of concrete was poured on top of the waste in each module, and a sand/gravel mixture was placed on top of the concrete roof. The dose rate at 1 m is approximately 10 mGy/h (1 mR/h).

This repository contains high- and medium-level waste, consisting of material from the roof of Unit 4, graphite, metal, concrete, and crushed stone. The exposure rate 1 m from this radioactive waste was 2.5 Gy/h (250 R/h) as of 1991. The waste is stored in containers and in bulk form. According to data provided by the Pripyat Dosimetric Monitoring Department, the total mass of waste at the site is 22,000 metric tons. The waste has a total activity of approximately $2.5 \times 10^{15}$ Bq, of which $1.85 \times 10^{15}$ Bq was in the form of $^{137}$Cs, approximately $7.4 \times 10^{14}$ Bq in the form of $^{90}$Sr, and $1.85 \times 10^{13}$ Bq in the form of $^{239,240}$Pu. The radioactive waste has a mean specific activity of $1.7 \times 10^8$ Bq. Other internal Ukrainian sources state the radioactivity is slightly higher, $2.6 \times 10^{15}$ Bq.

The disposal site is now partially conserved and equipped with monitoring wells as well as equipment for monitoring soil, groundwater, and surface atmospheric contamination. According to the Pripyat Dosimetric Monitoring Department, the surface contamination density in areas adjacent to the disposal facility is $7 \times 10^6$–$4 \times 10^7$ Bq/m$^2$ for $^{137}$Cs, $7 \times 10^6$–$5.5 \times 10^7$ Bq/m$^2$ for $^{90}$Sr, and $2 \times 10^5$–$2 \times 10^6$ Bq/m$^2$ for $^{239,240}$Pu. The groundwater level is 5–7 m below the surface. The $^{90}$Sr content of the groundwater ranges from 1.5–2 Bq/L, with some individual wells yielding 7 Bq/L. The atmospheric aerosol activity concentration in the layer adjacent to the ground ranges from $8 \times 10^{-4}$–$3.7 \times 10^{-2}$ Bq/m$^3$ (the maximum permissible concentration is $7.4 \times 10^2$ Bq/m$^3$).

### 4.1.10 Kompleksnyi Radioactive Disposal Site

This site is located in the incomplete low- and medium-level waste storage facility for the unfinished Chornobyl NPP Unit 5, 2.5 km southeast of Unit 4, on the east side of the cooling pond. The upper portion of the geological section consists of the an approximately 30-m-thick band of upper-Quaternary nonuniform alluvial sand, with interbedded loam and sand, underlain by a 20-m-thick artesian marl, clay, and aleurite Kiev formation horizon. Groundwater is at a depth of 3–4 m. The repository was constructed on a cast-in-place reinforced-concrete base and has no roof or end walls; only the top third of the building is above ground level.

The repository became full in 1987 and contains radioactive waste from the Chornobyl NPP, primarily in metal containers. The average filling factor is approximately equal to 0.7. Approximately 30% of the waste consists of metal, while the remainder consists of graphite, soil, and crushed stone. The radioactive waste was not accounted for upon receipt; however, the Pripyat Dosimetric Monitoring Department estimates that the total activity is approximately $1.3 \times 10^{15}$ Bq. The repository has a total area of 20,000 m$^2$. Various approaches for estimating the volume of radioactive waste housed in containers suggest that the actual volume of radioactive waste is on the order of 1,000 m$^3$.

### 4.1.11 **Rozsokha Equipment Holding Facilities I and 2**

The Rozsokha holding facilities, located within the 30-km Exclusion Zone, contain accident remediation equipment. Most of this equipment was used in April 1986 (after the accident) through November 1986 (after the completion of the Shelter).

Up until 1989, the process for mitigating the consequences of the accident was primarily managed by governmental agencies of the former Soviet Union. In 1990, military transport vehicles were removed from the balance sheets of particular military units and transferred for "storage" to the Chornobyl Auto-Transport Company. These vehicles were then moved onto a "parking" field near the village of Rozsokha, which is approximately 20 km southeast of the Chornobyl NPP. Currently, 1,621 vehicles are on the balance books of the state auto-transport enterprise "Chornobyl Lis." The surface contamination levels on these vehicles vary from 8.3 kBq/m$^2$ up to 6.6 x 10$^6$ Bq/m$^2$, and the collection of vehicles includes the following:

- 1,193 transport trucks

- 234 military vehicles

- 119 tow trucks

- 16 road construction vehicles

- 59 cars and buses.

In addition, 30 military helicopters are located in the main field and in a separate field about 1 km away—their air-intake nacelles are the primary contamination sites. The surface contamination levels of these helicopters vary from 1.6 kBq/m$^2$ up to 3.33 x 10$^{15}$ Bq/m$^2$.

The radiological conditions at Rozsokha at the end of April 1999 are characterized as follows:

- General background dose rate is 27 μR/hr.

- Ground-level concentration of $^{137}$Cs in the air is 3.6 x 10$^{-5}$ Bq/m$^3$; the control limit for this part of the 30-km Exclusion Zone is 3.7 x 10$^{-4}$ Bq/m$^3$.

- $^{137}$Cs deposition rate in the area is 0.16 Bq/(m$^2$ x day).

- Average surface contamination levels are 85 kBq/m$^2$ for $^{137}$Cs, 37 kBq/m$^2$ for $^{90}$Sr, and 0.48 kBq/m$^2$ for $^{238/239}$Pu.

- Groundwater contamination levels are 23 Bq/m$^3$ for $^{137}$Cs and 39 Bq/m$^3$ for $^{90}$Sr.

According to the Ministry of Emergency Situations, there is a plan to decontaminate the equipment located in the Exclusion Zone through a program called "Metal," but because of a lack of financing this has not been realized. Discussions were conducted between Ministry officials and representatives from Los Alamos National Laboratory.

For more information on this site and economic losses from contaminated equipment, see Section 8.1.2.

## 4.2 RADIONUCLIDE MIGRATION FROM INTERIM STORAGE AND DISPOSAL SITES

### 4.2.1 Radionuclide Migration and Worker Safety

Radionuclide migration does not pose a threat to workers at the interim radioactive waste storage and disposal sites. The primary risk to personnel working in or near the facilities (working on inventory, burial, research, etc.) is from external exposure. Inhalation hazards only arise when performing waste handling or trench remediation operations.

### 4.2.2 Radionuclide Migration and Public Safety

As far as the public living outside the Exclusion Zone is concerned, the burial sites do not pose any risk of external exposure or exposure via inhalation. The most likely impact involves leaching of radionuclides by surface water and groundwater into areas from which they can be discharged into rivers and streams (mainly the Pripyat River) and then into the Dnepr River basin, resulting in a risk of ingestion through drinking water or irrigation water pathways.

### 4.2.3 Radionuclide Migration to Monitoring Wells and Rivers

Table 4.2.1 lists the average radionuclide content in groundwater from monitoring wells adjacent to interim radioactive storage and disposal sites. The actual scatter in the values may run as high as 1000%.

There is still no consensus on the extent to which radionuclides from radioactive waste storage sites are reaching rivers and wells. Moreover, there is no basis for believing that the radionuclide content of the water in the monitoring wells corresponds to the level in the aquifer.

**TABLE 4.2.1.**  $^{137}$Cs and $^{90}$Sr in Groundwater Near Interim Radioactive Storage and Disposal Sites (Bq/L)

| Location of Monitoring Well | Aquifer | $^{137}$Cs | | | $^{90}$Sr | | | | |
|---|---|---|---|---|---|---|---|---|---|
| | | 1992 | 1993 | 1994 | 1992 | 1993 | 1994 | 1995 | 1996 |
| Podlesnyi RDS (Wells 4n, 5n 10n, 11n) | Quater. | 0.37 | 0.24 | 0.10 | 1.1 | 0.56 | 0.57 | 0.3 | 0.28 |
| Buryakovka RDS (Wells 5, 22, 35, 53, 65, 123) | Quater. | 0.78 | 0.21 | 0.17 | 1.0 | 1.2 | 0.51 | 0.36 | 0.3 |
| ChNPP Phase 3 RDS (Wells 4, 6, 7, 8, 13, 15) | Quater. | 0.92 | 0.56 | 0.25 | 1.9 | 1.5 | 1.1 | 0.55 | 0.6 |
| Ryzhii Les and Stroibaza IRSSs (Wells 1/1, 1/2, 2/1, 2/2, 2/3, 3/1) | Quater. | 0.33 | 0.44 | 0.18 | 34.0 | 27.4 | 59.7 | 63 | 59 |
| Ryzhii Les IRSS (Yanov Sta., Wells 516, 13, 14, 15) | Quater. | 0.22 | 0.18 | 0.49 | 0.96 | 0.52 | 0.89 | 0.7 | 0.6 |
| Ryzhii Les IRSS (Yanov Sta., Wells 515) | Quater. | 0.18 | 0.29 | 0.18 | 0.52 | 0.25 | 1.2 | 1.0 | 1.1 |
| Left-bank mouth (Krasnyankskaya, Wells 201-1, 201-2, 202-2, 203-1, 206-1, 206-2) | Quater. | 0.32 | 0.20 | 0.10 | 13 | 5.9 | 3.4 | 3.2 | 3.0 |

ChNPP = Chornobyl Nuclear Power Plant
IRSS = interim radioactive waste storage site
RDS = radioactive waste disposal site

Demchuk et al. [1990] and Voitsekhovitch et al. [1996b] predict that radionuclide migration from radioactive waste burial adjacent to the Pripyat River will be negligible over the next 50 years because the radionuclides will be retained in the soil and will not reach the groundwater. The total leachate of $^{90}$Sr from the major repositories will be 1.13 x 10$^{13}$ Bq/yr. The leachate from the Neftebaza and Peschanoe Plato into the Yanov Valley Bottom over the next decade will be 3.7 x 10$^{11}$–5.5 x 10$^{11}$ Bq/yr.

Based on the monitoring well data and surface water measurement made to date, we contend that radioactive waste burial in the Exclusion Zone will make virtually no additional contribution to contamination of water in the

Dnepr watershed. Surface water contamination is currently associated with radionuclide runoff from contaminated land (see Chapter 5). We do, however, believe that additional data is needed to confirm these preliminary conclusions.

# 5

ЭКОЛОГИЧЕСКИЙ

# ENVIRONMENTAL CONTAMINATION

V. Shestopalov and V. Poyarkov

## 5.1 SOIL CONTAMINATION IN THE EXCLUSION ZONE

The Exclusion Zone was established in 1986 following evacuation of the public and termination of economic activity near the Chornobyl Nuclear Power Plant (NPP). The Exclusion Zone is an area established around the accident site, which contains the highest contamination (see Figure 5.1.1 for more detail). The Exclusion Zone is roughly an area with a 30-km radius with the nuclear plant being at its center; it covers more than 2,044 km$^2$.

The Chornobyl NPP releases from 26 April–6 May 1986 contained several hundred different radionuclides, of which 20 were of high radiological significance (Table 2.5.1). Most of these, such as $^{131}$I, played an important role as a source of contamination during the first few days of the disaster and had short half-lives—days, weeks, or months. By 1987–1988, $^{137}$Cs (with a half-life of approximately 30 years) had become the predominant environmental contaminant both in terms of external radiation exposure and internal doses from ingestion of contaminated foods.

For personnel working in the Exclusion Zone and people living in areas within Ukraine contaminated by the accident (where the contamination is detectable, but is not high enough to require relocation), more than 90% of the external exposure received was because of $^{137}$Cs.

### 5.1.1 Contamination Map Construction and Data Analyses

The distribution of radionuclide contamination can be most effectively visualized using maps. Numerous attempts have been made to construct such maps, generally on the basis of unrelated data sets, which caused the maps to have substantial differences. The map in Figure 5.1.1, which is a

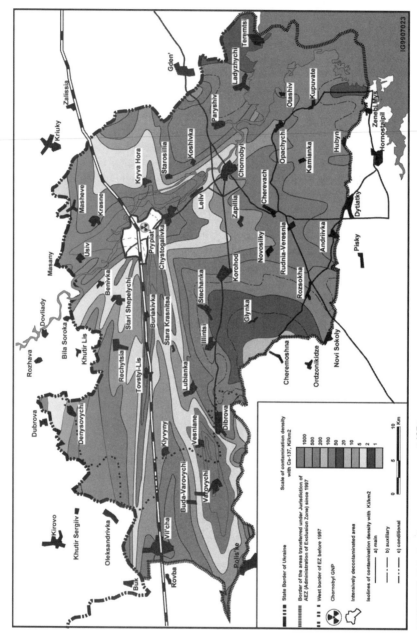

**FIGURE 5.1.1.** Summary map of $^{137}$Cs distribution within Exclusion Zone (as of 01 January 1996)

retrospective assessment of the $^{137}$Cs ground contamination as of 10 May 1986, was based on more than 40 different data sources obtained between 1986 and 1996.

The initial data consisted of a series of aerial gamma-ray profiles at scales of 1:100,000 and 1:25,000. In addition, numerous soil monitoring samples were collected from 1988–1996 over a "reference network." The reference network consists of a radially concentric network of monitoring points distributed along 36 radial lines (every 10°) running outward from Chornobyl NPP Unit 4. These points are located at radii of 5, 6, 7, 8.3, 10, 12, 14.5, 17, 20, 25, and 30 km from the plant. Further, direct gamma-ray spectrometric determinations of $^{137}$Cs in soil samples were available for nearly every population center in this area. Generally, from 5–30 analyses were performed per population center.

The major difficulties encountered in mapping the Exclusion Zone were as follows:

- The data sets do not provide for consistent, simultaneous sampling of the entire Exclusion Zone at sufficiently high density to draw maps to the desired scale (even at a level of 1–2 samples per square kilometer of the Exclusion Zone).

- Substantial differences exist in the basic mapping methods, sample collection methods, and laboratory analysis methods used with the samples.

- The soil analysis results are highly variable.

- The major migration processes are fairly severe, and quantitative indicators of surface migration are poorly understood. This makes it difficult to place data obtained at different times on a contemporaneous basis.

The available data were reduced to a "common denominator" using several principles, the most important of which are as follows:
- The statistical nature of all available single-point contamination density determinations was taken into account.
- Priority was given to the larger estimated radiation hazard whenever the data were inconsistent (this is based on the well-known conservatism principle from radiation protection).
- All structural dependencies, i.e., patterns in the locations of basic elements (such as "hot spots"), in the distribution of radionuclide contamination within the Exclusion Zone were identified.

Because of these difficulties, many Exclusion Zone maps showing the contamination from various radionuclides are extremely low in accuracy. Table 5.1.1 compares the estimated mean $^{137}$Cs contamination densities for various population centers within the Exclusion Zone given by seven different data sources, including both primary references and cartographic materials. Approximately a third of the values show strong deviations (a factor of 2–5) relative to the standard (modal) values.

### 5.1.2  Initial $^{137}$Cs Distribution in Exclusion Zone

Analysis of the extensive data used in constructing the Exclusion Zone contamination map for cesium indicates four distinct types of contamination fields (Figure 5.1.1):

- Centralized, concentric structure: contamination levels increase toward the epicenter of the disaster, Chornobyl NPP Unit 4
- Linear/radial structures: generally narrow areas of elevated contamination levels called "plumes," which are frequently divided into second-order structures ("jets").
- Local structures containing elevated contamination levels: oval areas not connected to the central concentric structure
- Relatively uncontaminated "wedges" between major plume: areas generally expand toward the periphery of the Exclusion Zone.

In addition to mapping the Exclusion Zone, we also created contamination maps for areas beyond the Exclusion Zone boundary. This enabled us to more accurately determine the structure of the contamination field within the zone.

Our map (Figure 5.1.1) shows a broad range of contamination levels, from less than 18 kBq/m$^2$ to more than 3.7 x 10$^7$ Bq/m$^2$. From 1–3 km from Unit 4, the calculated $^{137}$Cs densities in May 1986 ranged from 1.8 x 10$^8$ Bq/m$^2$–7.4 x 10$^8$ Bq/m$^2$.

Several plumes are visible on Figure 5.1.1 radially diverging from Chornobyl NPP. The initial release from the explosion on 26 April 1986 led to the deposition of a very narrow fuel plume toward the west (with radioactive material also being deposited on 27 April). The West Plume has the highest contamination level. This plume is an extremely narrow, nearly linear band only 1–5 km wide and up to 80 km in length. The contamination levels decrease by a factor of 5–10 within only 1–1.5 km from center to edge on both the north and south sides of the plume. The contamination level in the West Plume has a value of 1.8 x 10$^6$–2.6 x 10$^6$ Bq/m$^2$ where it crosses the boundary of the Exclusion Zone. Within Ukraine, this plume covers Kiev and Zhitomir Oblasts, as well as the northern portion of Rivno Oblast and the northeast portion of Volynsk Oblast.

**TABLE 5.1.1.** Comparison of Estimated Mean $^{137}$Cs Contamination Levels (kBq/m$^2$) for Population Centers in the Exclusion Zone

| Population Center | BES-4[a] | BES-5[b] | BD+DP[c] | Aerial Gamma-Ray Spectroscopic Maps[d] | | Maps Based on Individual Soil Samples[e] | |
|---|---|---|---|---|---|---|---|
| | | | | 1:100,000 | 1:25,000 | Science and Technology Center (1994) | International Science and Technology Center (1994) |
| Pripyat | 777–3293 | — | — | 1850–7400 | 1850–4440 | 3700–7400 | |
| Chornobyl | 333–814 | — | 591.6 | 185–814 | 555–1110 | 370–1110 | 259–444 |
| Chistogalovka | 1443 | 9250 | 823.6 | 1480–4440 | 740–3700 | 3700–7400 | 740–1480 |
| Dityatki | 11.1 | 144.3 | 151.7 | >37 | 148–222 | 148–222 | 0–185 |
| Kopachi | 666 | 129.5 | 5583 | 5550–11000 | 1110–3700 | 3700–7400 | 2590–5550 |
| Buryakovka | 10730 | 4810 | 114.5 | 3700–14800 | 1850–14800 | 3700–11100 | 7400–14800 |
| Korogod | 34.04 | 203.5 | 423.7 | 18.5–74 | 111–185 | 148–222 | 111–222 |
| Stechanka | 85.1 | 48.1 | 78.1 | 18.5 | 29.6–81.4 | 74–111 | 0–185 |
| Shepelichi Sta. | 481 | 477.3 | 788.1 | 925–1480 | 555–1110 | 740–1480 | 555–1480 |
| Zelenyi Mys | 7.4–24.1 | 40.7 | 97.7 | >37 | 74–148 | 74–148 | |
| Opachichi | 7.4–362.6 | 214.6 | 999.4 | 74–370 | 185–370 | 185–296 | 148–222 |
| Rudnya Il'inskaya | 74–114.7 | 129.5 | 447.7 | 18.5–185 | 37–185 | 55.5–111 | 0–185 |
| Otashev | 74–88.8 | 358.9 | 592.7 | >37 | | 74–148 | 111–185 |
| Zalesye | 133.2–703 | 407 | 334.8 | 111–592 | 222–444 | 444–740 | 185–370 |
| Illintsy | 18.5–107.3 | 66.6 | 71.8 | 18.5 | 37–259 | 55.5–148 | 0–185 |

(a) Data published in *Byulleten Ekologicheskaya Sostoyanie Zony Otchuzhdeniya*, No. 4.
(b) Data published in *Byulleten Ekologicheskaya Sostoyanie Zony Otchuzhdeniya*, No. 5.
(c) Database maintained by Radiation Protection and Chornobyl NPP Dosimetric Assessment Department, Ukraine.
(d) Aerial gamma-ray spectrometric maps obtained in 1989.
(e) Data obtained for reference network (1992–1994) (ISTC (96)–Shelter International Science and Technology Center data obtained in 1996).

The widest of the plumes, the fan-shaped South Plume, runs south-south-east of Chornobyl NPP. This plume—which has the second highest contamination level—passes through Kiev, Cherkassy, and Kirovograd Oblasts, as well as portions of Vinnitska, Odessa, and Nikolaevsk Oblasts. In addition, this plume includes a spotty band of contamination that passes through portions of Vinnitsa, Khmel'nitskii, Ternopol', Ivano-Frankovsk, and Chernovits Oblasts. The South Plume gradually divides into 4–5 separate jets, which have a higher longitudinal gradient in the contamination level than the North and West Plumes. The $^{137}$Cs densities rapidly decrease from 7.4 x $10^6$ -18 x $10^6$ Bq/m$^2$ near Chornobyl NPP to 1.8 x $10^5$–3.7 x $10^5$ Bq/m$^2$ at the southern border of the Exclusion Zone. The contamination levels in the gaps between the individual jets decrease to 1.1 x $10^5$–1.8 x $10^5$ Bq/m$^2$.

The North Plume runs to the northwest of Chornobyl NPP at an azimuth of approximately 340°. This contamination plume was released during the fire subsequent to the explosion. It is wider than the West Plume (as much as 6–10 km wide) but much shorter. At 13–15 km from the power plant, the North Plume merges into the Radinsko-Kryukovskaya structure, which covers almost the entire northern half of the 30-km zone and extends into Belarus. The contamination levels on the border with Belarus from the North Plume ranged from 7.4 x $10^6$–15 x $10^6$ Bq/m$^2$.

The East Plume, Northwest Plume, and Southwest Plume each consist of 1–2 individual jets of radioactive contamination. The first two of these three plumes die out some 20–30 km from the plant. The Southwest Plume regains strength at 30–32 km from the plant, where the cesium contamination levels rapidly increase from 3.7 x $10^5$–7.4 x $10^5$ Bq/m$^2$ near the village of Lubyanka to 1.8 x $10^7$–3.3 x $10^7$ Bq/m$^2$ among the villages of Vesnyanoe, Dibrova, and Martynovichi. At the border of the Exclusion Zone, the Polessko-Narodichskii Plume—the most-contaminated area in Ukraine outside the Exclusion Zone—begins. The Polessko-Narodichskii Plume had an initial $^{137}$Cs contamination greater than 3.7 x $10^6$ Bq/m$^2$ and extends for 60 km, passing over Vesnyanoe, Polessko, Shishelovka, and Bober. Only the first few kilometers of this band lie within the Exclusion Zone.

In addition to the Vesnyanoe structure (which is 2–4 km wide), there is an extensive area of high contamination in the northern portion of the Exclusion Zone. This high contamination area in the north, bounded by the 7.4 x $10^6$ Bq/m$^2$ contour, is up to 40 km in length and 8–10 km in width. The contamination levels at the center of this anomaly (near the villages of Radin, Kulazhin, and Kryuki) reach from 1.8 x $10^7$–7.4 x $10^7$ Bq/m$^2$.

### 5.1.3 **Redistribution of $^{137}$Cs Contamination Since the Accident**

The basic difference between the current Exclusion Zone contamination and that of May 1986 is that the activity has become buried over the bulk of the territory, and has also decreased as a result of radioactive decay. The $^{137}$Cs activity in the top 10–20 cm of soil usually analyzed for radiation monitoring and mapping has generally decreased by 20–50% in the 10 years following the accident. This decrease is largely due to the following processes:

- Natural radioactive decay of $^{137}$Cs; the annual rate of decay is approximately 2.3%

- Surface runoff, anywhere from 0–30% per decade, depending on slope, soil type, and vegetation (based on several expert opinions)

- Decontamination of various areas and population centers within the Exclusion Zone

- Vertical migration of radionuclides within the soil.

The reductions in radionuclide contamination were most substantial immediately adjacent to Chornobyl NPP, i.e., within 2–5 km. A significant fraction of this area (the Chornobyl NPP site, the city of Pripyat, the Ryzhii Les area, and several radioactive waste burial areas) underwent fairly intensive surface decontamination. Decontamination involved two activities: removing radioactive fallout and burying fallout under relatively uncontaminated soil. Decontamination reduced the mean $^{137}$Cs surface contamination levels by a factor of 10–100 (more in some areas) relative to the 1986 contamination levels.

On the other hand, secondary (also known as local) accumulation of radionuclides has occurred in the surface layer in small areas of the Exclusion Zone (on the order of 5–10% of the area). The secondary accumulation has occurred in low-lying areas that receive material washed from surrounding high areas and slopes during spring snowmelt and rainstorms. These low-lying areas include the edges of swamps, minor sinkholes and depressions, ravine bottoms and alluvial fans, and other similar relief forms. Experimental observations suggest that surface contamination levels in such areas could increase by a factor of 2–5 over a 10-year period, as confirmed by our Belorussian colleagues at the Masany experimental station 12 km from Chornobyl NPP.

The following conclusions for radioactive contamination levels in the Exclusion Zone have been reached:

- The initial surface contamination levels in the Exclusion Zone vary widely: from 3.7 x $10^3$ Bq/m$^2$–7.4 x $10^8$ Bq/m$^2$ with rates being higher at the Chornobyl NPP site itself. The contamination field generally has a radial/jet structure.

- The radiation environment is relatively stable over most of the Exclusion Zone and is self-cleansing at a rate of approximately 2–5% per year.[1]

- The radiation environment and radioenvironmental conditions may deteriorate as a result of increased radionuclide contamination. This deterioration may be observed in localized areas where the surface landscape and geomorphological conditions favor secondary accumulation of material washed from adjacent areas.

### 5.1.4 Surface Contamination in the Exclusion Zone from $^{90}$Sr and Transuranic Radionuclides

Very few summary contamination maps exist for $^{90}$Sr and transuranic elements such as $^{238}$Pu, $^{239+240}$Pu, and $^{241}$Am released by the Chornobyl NPP accident. These maps were not made for several reasons. These radionuclides cannot be measured directly, i.e., from aircraft or helicopter; thus, the contamination from these radioisotopes was measured by soil analyses. Because soil samples were used, a much smaller standard volume of samples was acquired; hence, the samples are not highly representative, increasing the variability of the measurements. This variability gives rise to serious additional difficulties when constructing maps, and radically increases the ambiguity and subjectivity of contours. The soil analyses were quite expensive (tens to hundreds of times more expensive than gamma-ray spectrometric analysis of the same samples for $^{137}$Cs and other gamma emitters), and thus fewer analyses were conducted. Further, the weighed soil samples were very heterogeneous because the $^{90}$Sr and transuranic-element isotopes are almost entirely associated with the fuel particles and do not form condensed fallout, which provides greater sample homogeneity. For these reasons, maps of the $^{90}$Sr and transuranic-element contamination in the Exclusion Zone were published in 1996 (10 years after the accident). These maps have 10 contamination-density levels, rather than 2–4 levels found on the State Committee for Hydrology and Meteorology maps. The contamination density levels allowed us to analyze the internal structure of the contamination field and estimate maximum contamination levels and gradients in the contamination field. With these maps, we could analyze the contamination

---

[1]Editor's Note: Of this, approximately 2.3% per year is the result of physical decay of $^{137}$Cs. The balance is the result of environmental transport processes.

fields more thoroughly and with a better foundation than allowed by any previously published materials.

In 1989 and 1990, soil samples were collected from the Exclusion Zone. These samples were analyzed for $^{90}$Sr and transuranic elements in an effort to construct contamination maps [Shestopalov 1996]. The sample collection points were co-located with points in the so-called "reference network," with approximately 340–350 points in the Exclusion Zone. However, the insufficient number of sampling points and the nonuniform distribution of reference network points prevented us from actually constructing the desired maps. Maps were constructed somewhat later by the Pripyat Science and Technology Center; however, the resulting maps have serious deficiencies, both for the reasons listed above and because of inadequacies in the map-compilation techniques in common use at that time.

In 1997 under a task order issued by the Exclusion Zone Administration, soil samples were collected from a regularly spaced grid. The grid had a mesh size of approximately 1 km (samples were collected on a grid of 100–500 m in areas where there was a large gradient in the contamination levels) to construct a more reliable map. The coordinates of the sampling points were determined using a global positioning system. The soil samples (five were collected at each sampling point) were collected to a depth of 30 cm. Completeness of $^{90}$Sr collection was monitored using the vertical distribution of $^{90}$Sr and the activity ratio of $^{90}$Sr and $^{154}$Eu in each sample. The sample activities were used to construct 1:200,000-scale maps of the $^{90}$Sr contamination and the $^{137}$Cs/$^{90}$Sr ratio as of 1997. As the existing large-scale aerial gamma-ray spectrometric maps of the $^{137}$Cs contamination density within the Exclusion Zone were already well integrated, these maps were combined with the $^{137}$Cs/$^{90}$Sr ratio soil-sampling maps to construct a map of the $^{90}$Sr contamination in the zone [UNIISKhR 1998].

The $^{90}$Sr contamination field is generally well correlated with the $^{137}$Cs contamination. The highest contamination is observed at Tolstyi Les in the West Plume (the "narrow" plume). The maximum contamination levels, which occur 2–5 km from Chornobyl NPP, are $1.8 \times 10^7$–$3.7 \times 10^7$ Bq/m$^2$, and decrease to $0.4 \times 10^6$–$1.5 \times 10^6$ Bq/m$^2$ at the boundary of the 30-km zone. The following $^{90}$Sr contamination levels are observed: Chornobyl NPP—approximately $0.4 \times 10^6$ Bq/m$^2$; along most of the periphery of the 30-km zone—37–180 kBq/m$^2$; and along the northern boundary of the 30-km zone—180–370 kBq/m$^2$. The contamination levels for plutonium and $^{241}$Am are tightly correlated with those for $^{90}$Sr.

The maximum concentrations of the transuranic elements coincide with the maximum concentrations of $^{90}$Sr, with values as follows: approximately $7.4 \times 10^5$ Bq/m$^2$ for the two plutonium isotopes and approximately $5.5 \times 10^5$ Bq/m$^2$ for $^{241}$Am. These concentrations occur within the Exclusion

Zone. At the periphery of the 30-km zone, these values decrease to 1.8 kBq/m$^2$ or even lower. In addition to the West Plume, the wide North Plume is prominent on all of the maps, with 2–3 jets. The less active South Plume has 2–3 separate jets, one of which follows the Pripyat River flood-plain just to the east of the city of Chornobyl. The overall pattern for transuranic elements closely resembles the $^{137}$Cs contamination field described in Sections 5.1.2 and 5.1.3.

### 5.1.5 Estimated Total Activities for the Main Radionuclides in the Exclusion Zone

The substantial amount of primary data available describing the environmental contamination in the Exclusion Zone naturally creates the impression that the total activity contained in this area must have been determined to a high accuracy long ago. Critical analysis of the actual data unfortunately demonstrates this is not so. Extensive usage of modern systems-oriented methods for the analysis of various data, as well as modeling the composition of the radionuclide fallout and the laws governing the spatial variation in contamination level and composition, may enable the accuracy of the information to be improved.

In our opinion, the most important thing is to generate some internal consistency among various integrated data sets on total radionuclides using a systems-based or structured approach. Three data sets are of concern: the radionuclides over the surface of the 5-km zone around Chornobyl NPP; radionuclides in the radioactive waste in interim radioactive waste storage and disposal sites; and the ratios of activity for various radionuclides in the initial fallout in spring 1986 and long-lived radionuclides currently present.

Making this data more consistent will not only serve as an example of the application of the well-known systems approach to analysis of accumulated data, but also as a unique internal self control and a method for increasing the accuracy and validity of overall estimates for the scale of radioactive contamination in the Exclusion Zone. Speaking more specifically, the accuracy of these integrated indicators can be increased through the following measures:

- Performing combination and spatial analysis of existing differential contamination-level data

- Obtaining estimated totals for individual radionuclides through numerical integration of the appropriate contour maps

- Correcting the data for various radionuclides (this includes determining their ratios; intercalibrating with respect to area, direction of release, and type of precipitation-related fallout; and calibrating against the initial fuel composition fixed in various types of fallout).

The change in total radionuclide activity when the calculation method is changed to make the data more consistent could be as much as 30–50% or even more. However, it is well known that when the gradient fields are complex and the sampling network is nonuniform, it is in principle more accurate to determine the total activity from contour maps than by using the "arithmetic mean" sample.

### 5.1.5.1 Total Cesium Activity in the Exclusion Zone (Excluding Chornobyl NPP).

All preliminary determinations discussed in this section are based on the 1:100,000-scale summary map of $^{137}$Cs contamination levels in the Exclusion Zone as of May 1986 (see Figure 5.1.1).

The basic features of this map are as follows:

- A wide variety of input data was used in constructing the map (more than 40 sources)

- The input data were combined using the new structured approach for construction of radionuclide contamination maps mentioned in Section 5.1.5

- Random noise was filtered out, i.e., the actual representativeness of the input data was taken into account, when drawing the contour lines

- The overall curvature of the surface of the radionuclide-concentration field was taken into account when determining the values for a typical section

- The contour lines were closely spaced—up to 12–14 contamination levels were used, instead of the standard 5 levels used previously.

Using this map, the estimated total amount of $^{137}$Cs that initially fell on the Exclusion Zone is much more accurate than any non-map-based method for combining the input data sets, either using other maps compiled without previous analysis of the contamination-field structure or on the basis of more limited input data. Thus, the map in Figure 5.1.1 was used to obtain an approximate estimate for the numerically integrated total amount of $^{137}$Cs deposited over the entire Exclusion Zone except for the Chornobyl NPP site itself. A different contamination mechanism based on the scattering of ejecta from the explosion, rather than atmospheric deposition of finely dispersed fractions, was used for the NPP site.

The system for obtaining the estimates of various total contamination indicators is based on the following criteria, preconditions, and assumptions:

- All radioactive waste in disposal sites is treated as being at the Chornobyl NPP site.

- Half of the radioactive waste in the Stroibaza Interim Radioactive Waste Storage Site and all other interim radioactive waste storage sites are treated as lying in areas beyond the Chornobyl NPP site.

- Reasonably reliable, standard ratios between the activities of the major radionuclides are used, both for the initial fuel and for contamination within the Exclusion Zone (the two basic types of radionuclide fallout—fuel and condensation—were taken into account)

- Recent data on the volume of radioactive waste in interim storage sites (see Chapter 4) were used, breaking down the activities by major radionuclides using the same criteria as those used for the area surrounding the interim radioactive waste storage sites in question.

- Ambiguities in data were resolved using the conservativeness principle, i.e., the more pessimistic values were used.

In addition, any change in total activity between 1986 and 1997 was assumed to be the result of radioactive decay alone. Radionuclide transport outside the Exclusion Zone amounts to no more than 1–3% of the total. The fraction of $^{137}$Cs (and, more importantly, $^{90}$Sr)[2] that has migrated below the 20-cm level is unknown. In any case, this portion of the activity remained within the Exclusion Zone. The activity of $^{241}$Pu was included in the total activity of a repository (traditionally the total activity in a repository is quoted without $^{241}$Pu, which has in recent years come to represent approximately 15% of the total activity produced by all radionuclides).

Factoring in these criteria, preconditions, and assumptions with the analyzed data leads to the following estimated total values for $^{137}$Cs:

- In 1986, the initial total value of fallout for the Exclusion Zone (not including Chornobyl NPP site) was 7.2 x $10^{15}$–8.1 x $10^{15}$ Bq; a mean value of 7.8 x $10^{15}$ was adopted

- In 1997, the total value of fallout for the Exclusion Zone was 6.3 x $10^{15}$ (taking radioactive decay into account)

- $^{137}$Cs makes up an average of 60% of the radioactive waste in the interim radioactive waste storage sites as implied by the standard value of the ratio $^{137}$Cs/$^{90}$Sr = 1.2–1.7 (with a mean value of approximately 1.5); suggesting that the total activity of $^{137}$Cs in the interim radioactive waste storage sites is 0.7 x $10^{15}$ Bq

---

[2]Editor's Note: The environmental transport of $^{90}$Sr is of particular concern because it is a bone-seeking radionuclide and, therefore, has a much higher radiotoxicity per unit intake than $^{137}$Cs.

- The total activity of $^{137}$Cs in the Exclusion Zone (not including the interim radioactive waste storage sites) is 5.6 x 10$^{15}$ Bq

- In 1997, approximately 0.16 x 10$^{15}$ of $^{137}$Cs (taking decay into account) is in deposits at the bottom of the cooling pond.

**5.1.5.2 Total Strontium Activities in the Exclusion Zone (Excluding Chornobyl NPP).** We determined the total $^{90}$Sr in 1986 and 1997 within the Exclusion Zone using an indirect method based on the $^{137}$Cs/$^{90}$Sr ratio and the variations in this ratio as a function of position within the Exclusion Zone. The average cesium/strontium ratio for spent fuel is 1.4, for the overall atmospheric release from the Chornobyl NPP accident is 8.5 (as of 1996), and for hot fuel particles is 0.8. Combining the results of Exclusion Zone soil analyses performed by the Chornobyl Center of International Research, Ukrainian Academy of Sciences Institute for Nuclear Research, State Enterprise for Radiation, Dosimetric and Environmental Monitoring, and other organizations indicated that the mean $^{137}$Cs/$^{90}$Sr ratio varies. Quite systematic variations are seen from 1.2–1.7 within 2–5 km of Chornobyl NPP (or even 0.8–1.5 at the Chornobyl NPP site itself) to 2.5–4.5 at the periphery of the Exclusion Zone, increasing to 20–50 or even higher in areas impacted by atmospheric fallout. This includes large areas of Belarus (the Radin-Kulazhin-Kryuki area in the northern half of the Exclusion Zone) as well as portions of the Exclusion Zone (in Ukraine)—the Denisovichi area and the Vesnyanoe-Dibrova area. Thus, a standard cesium/strontium ratio of 1.5–3.0 with an area-weighted mean value of 2.0 for the Exclusion Zone was adopted.

Using this information and the initial total of 7.8 x 10$^{15}$ Bq for $^{137}$Cs (Section 5.1.5.1), the following values are estimated for $^{90}$Sr in the Exclusion Zone:

- In 1986, the initial total fallout for $^{90}$Sr within the Exclusion Zone was 3.9 x 10$^{15}$ Bq (continuing to ignore the high-level radioactive waste in the vicinity of the Chornobyl NPP site and the high-level radioactive waste in the interim radioactive waste storage sites)

- In 1997, the total amount of $^{90}$Sr was 3.1x 10$^{15}$ Bq (taking radioactive decay into account)

- In 1997, the total amount of $^{90}$Sr in interim radioactive waste storage sites was 0.5 x 10$^{15}$ Bq

- In 1997, the total amount of surface $^{90}$Sr in the Exclusion Zone (not including interim radioactive waste storage sites or Chornobyl NPP site) was 2.6 x 10$^{15}$ Bq

- In 1997, total deposit of $^{90}$Sr at bottom of cooling pond was 0.1 x 10$^{15}$ Bq (using a mean cesium/strontium ratio of 1.6 for this area of the Exclusion Zone).

### 5.1.5.3 Total Plutonium in the Exclusion Zone (excluding Chornobyl NPP). Using a scheme similar to that used to calculate $^{90}$Sr, we calculated the total amounts of the two main plutonium isotopes, $^{239}$Pu and $^{240}$Pu. The standard strontium:plutonium ratios (for $^{239}$Pu and $^{240}$Pu) are as follows [Ministry for Chornobyl Affairs 1996]:

- 100:1 for Chornobyl NPP Unit 4 fuel
- 140:1 for entire accident release
- 32:1 for Chornobyl NPP 5-km zone (as of 1995)
- 60:1 for Chornobyl NPP 5-km zone (as of 1992) and fuel component of release [N.P. Arkhipov 1994].

This ratio varies by as much as a factor of 2–4, while the time factor only has a 10–20% effect.

The results quoted by *Role of Natural and Man-Made Factors in Radionuclide Migration through Soil and Vegetation in Various Areas* [N.P. Arkhipov 1994] are the most trustworthy because they represent the most comprehensive analysis of data available on the subject. These results include a correction for the amount of $^{90}$Sr that decayed during the 6 years since the study as well as the fact that the strontium:plutonium ratio increases slightly with distance from Chornobyl NPP. The estimated strontium:plutonium ratio for the Exclusion Zone as a whole as of 1986 is 70:1, implying the following $^{239,240}$Pu totals:

- Initial $^{239,240}$Pu fallout in Exclusion Zone was approximately 5.6 x 10$^{13}$ Bq (which remains virtually constant, given the long half-lives of these radionuclides)

- Amount of plutonium in interim radioactive waste storage sites is 0.7 x 10$^{13}$ Bq

- Total amount of these two plutonium isotopes on the Exclusion Zone surface (not including interim radioactive waste storage sites or Chornobyl NPP site) is 4.9 x 10$^{13}$ Bq

- Total plutonium deposits at the bottom of cooling pond is 0.14 x 10$^{13}$ Bq.

### 5.1.5.4 Total Alpha and Beta Emitter Activities in the Exclusion Zone (Excluding Chornobyl NPP). There is a certain amount of interest in

estimating analogous amounts for several transuranic elements—the $\alpha$-emitters $^{238}$Pu and $^{241}$Am and the $\beta$-emitter $^{241}$Pu. These estimates may be derived using the ratios relative to the main plutonium isotopes ($^{239,240}$Pu).

For $^{238}$Pu, the $^{238}$Pu:$^{239,240}$Pu ratio is relatively stable [N.P. Arkhipov 1994] and varies from 0.50–0.55, leading to a total estimated activity of 2.7 x $10^{13}$ Bq (upper limit) for $^{238}$Pu. Very little experimental material is available for $^{241}$Pu. Thus, the estimated total activity is based on the ratios relative to $^{239,240}$Pu in the initial fuel, taking the relatively short half-life (14.2 years) into account. As of 1997, $^{241}$Pu: $^{239,240}$Pu ratio was approximately 50:1; as of 2026, it will be 12.4:1; and by 100 years after the 1986 accident, it will have decreased to a value less than unity (0.7). Thus, the total activity of $^{241}$Pu is approximately 2.5 x $10^{15}$ Bq; however, this value will decrease by roughly a factor of 4 over the next 30 years.

The situation is slightly more complex for $^{241}$Am, a daughter product from the decay of $^{241}$Pu. The relatively long half-life of americium (approximately 433 years) relative to the half-life for $^{241}$Pu leads to the well-known secondary accumulation effect, which will continue over the next several decades. In 1997, the $^{241}$Am:$^{239,240}$Pu ratio was approximately 1.2; by the middle of the 21st century, it will increase to approximately 2.6–2.9, and then begin to slowly decrease. The total activity of $^{241}$Am, which is currently approximately 5.9 x $10^{13}$ Bq, will behave similarly.

**5.1.5.5 Conclusions Regarding Activities in the Exclusion Zone.** As of 1997, the following values are recommended for the estimated total activities of the radionuclides distributed over the surface of the Exclusion Zone and at the bottoms of bodies of water within the zone (not including the Chornobyl NPP site):

- $^{137}$Cs     5.6 x $10^{15}$ Bq
- $^{90}$Sr     2.6 x $10^{15}$ Bq
- $^{239,240}$Pu     4.9 x $10^{13}$ Bq.

These values and the more detailed data in Table 5.1.2 are based on careful analysis of the data and a highly accurate map of the $^{137}$Cs distribution. Further, the analyses took into account that the values were measured at different times (1986–1997). Finally, the total activities of the radioactive waste in the interim radioactive waste storage sites were included. This by no means implies that the total radioactive contamination has been determined. To the contrary, these results should be considered a first, preliminary step. The goal of these calculations and comparisons is to show that the previous values could and should be improved using an integrated approach to data summarization.

**TABLE 5.1.2.** Estimated Total Activities of Radionuclides in Exclusion Zone, 1997 (Bq)[a]

| Radionuclide | Radionuclides in Repositories[b][c][d] | | Total Surface Activity for Exclusion Zone | | Total |
| | Interim Radioactive Waste Disposal Site | Bottom Radioactive Waste Storage Sites | Deposits in Cooling Pond | Total | |
| --- | --- | --- | --- | --- | --- |
| $^{137}$Cs | 3.4 x 10$^{15}$ | 1.1 x 10$^{15}$ | 0.16 x 10$^{15}$ | 5.6 x 10$^{15}$ | 10.1 x 10$^{15}$ |
| $^{90}$Sr | 2.8 x 10$^{15}$ | 0.7 x 10$^{15}$ | 0.1 x 10$^{15}$ | 2.6 x 10$^{15}$ | 6.1 x 10$^{15}$ |
| $^{241}$Pu | 2.7 x 10$^{15}$ | 0.7 x 10$^{15}$ | 0.1 x 10$^{15}$ | 2.5 x 10$^{15}$ | 5.9 x 10$^{15}$ |
| Transuranic $\alpha$-emitters ($^{238}$Pu + $^{239,240}$Pu + $^{241}$Am [e] | 1.4 x 10$^{14}$ / 2.3 x 10$^{14}$ | 0.4 x 10$^{14}$ / 0.6 x 10$^{14}$ | 5.0 x 10$^{12}$ / 8.5 x 10$^{12}$ | 1.3 x 10$^{14}$ / 2.2 x 10$^{14}$ | 3.1 x 10$^{14}$ / 5.1 x 10$^{14}$ |

(a) All numbers rounded to two significant digits.
(b) Total activities of major radionuclides in radioactive waste disposal sites and interim radioactive waste storage sites from Table 4.1.1, with data for $^{241}$Pu, $^{238}$Pu, and $^{241}$Am added.
(c) All of the activity in interim radioactive waste storage sites, including all activity in the Stroibaza Interim Radioactive Waste Storage Site, is included.
(d) The numerator in the $\alpha$-emitter values is for 1997, while the denominator corresponds to the maximum activity (due to the increase in $^{241}$Am activity) in the 21st century.
(e) The mean cesium:strontium ratio for the radioactive waste in the disposal sites was assumed to be 1.2, while that for the radioactive waste in interim storage sites was assumed to be 1.5, consistent with the values for the Chornobyl NPP site and the area within 5 km.

## 5.2 SOIL CONTAMINATION OUTSIDE THE EXCLUSION ZONE

### 5.2.1 Soil Contamination in Europe

Almost two thirds of the cesium released by the Chornobyl NPP accident is concentrated in Belarus, Russia, and Ukraine. However, because the air masses containing the radioactive materials moved over the Northern Hemisphere for several weeks, almost every European country received some contamination, with the Scandinavian countries and countries in the Alps being more severely affected than others. Approximately 10% of the total release, or 8 x 10$^{15}$ Bq, is in the Scandinavian countries. The radioactive contamination fields outside the former Soviet Union were generally shaped during the night of 27 April 1986 and through the next 10 days. Rainfall during this period caused increased fallout in Sweden, Finland, Germany, Austria, Switzerland, Greece, Bulgaria, Romania, and Georgia.

Table 5.2.1 describes the $^{137}$Cs distribution in Europe and provides some insight into the characteristics of the radionuclide distribution as a function of distance from Chornobyl NPP.

Figures 5.2.1 and 5.2.2 clearly show a second peak in the distribution of total $^{137}$Cs activity at 1,000–1,400 km from Chornobyl NPP; this corresponds to the elevated contamination levels in the Alps, Balkans, and Finland. There is another local maximum associated with anomalous contamination in Sweden and Norway at a distance of 1,500–1,900 km.

Mountain ranges and plateaus, which received unusually high radionuclide fallout at distances more than 500 km from Chornobyl NPP, played an important role in determining the distribution of radioactive contamination throughout Europe. The Carpathian foothills, the Donetsk Hills, the

**TABLE 5.2.1.** Differential and Integral Distribution of $^{137}$Cs in Europe

| Inner Radius of Contamination Zone ($R_{in}$), km | Outer Radius of Contamination Zone ($R_{out}$), km | Area Within the Zone as a Percentage of the Total Area of Europe (S), % | $^{137}$Cs Activity Within a Specific Zone as a Percentage of the Entire Zone's Contamination (Q), % | $^{137}$Cs Contamination Level (q), kBq/m$^2$ |
|---|---|---|---|---|
| 0 | 10 | 0.0034 | 1.70 | 5030 |
| 10 | 30 | 0.0275 | 4.69 | 1730 |
| 30 | 100 | 0.3129 | 7.19 | 235 |
| 100 | 400 | 5.1587 | 24.11 | 48 |
| 400 | 800 | 15.275 | 16.49 | 11 |
| 800 | 1400 | 30.176 | 25.46 | 8.6 |
| 1400 | 2000 | 32.695 | 15.47 | 5.7 |
| 2000 | 3000 | 16.355 | 4.89 | 3.1 |
| 0 | 10 | 0.003 | 1.70 | 5030 |
| 0 | 30 | 0.03 | 6.39 | 2100 |
| 0 | 100 | 0.34 | 13.58 | 400 |
| 0 | 400 | 5.50 | 37.69 | 70 |
| 0 | 800 | 20.78 | 54.18 | 27 |
| 0 | 1400 | 50.95 | 79.64 | 16 |
| 0 | 2000 | 83.65 | 95.11 | 12 |
| 0 | 3000 | 100 | 100 | 10 |
| 800 | 3000 | 79.226 | 45.82 | 6.0 |
| 1400 | 3000 | 49.05 | 20.36 | 4.2 |

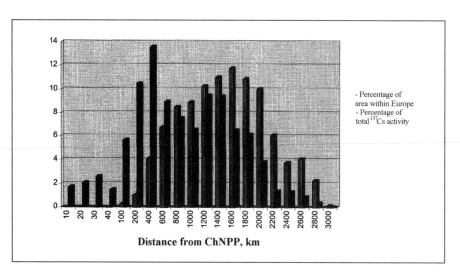

**FIGURE 5.2.1.** Percentage of area within Europe (gray) and amount of $^{137}$Cs as a function of distance from Chornobyl NPP (black) (% of total in each case)

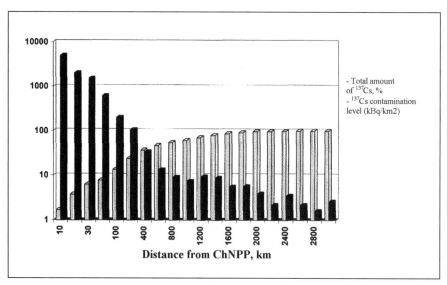

**FIGURE 5.2.2.** Total percentage of $^{137}$Cs (gray) and $^{137}$Cs contamination level (kBq/km$^2$) (black) as a function of distance from Chornobyl

Appennines, the Alps and Balkans (as mentioned above), the Scottish High-lands, and other mountains and plateaus had small, 20- to 100-km$^2$ areas with contamination levels of 100 kBq/m$^2$ or more. Threshold anomalies with $^{137}$Cs densities greater than 200 kBq/m$^2$ were detected in only one West European country—England, but these turned out to be unrelated to the acci-dent and were instead a result of local nuclear enterprises.

The situation throughout Europe can briefly be described as follows: the accident at Chornobyl NPP left its mark on over two thirds of the continent, and $^{137}$Cs contamination levels are now higher than they were before the accident.

### 5.2.2 Radioactive Contamination of Ukraine

The accident at Chornobyl NPP significantly changed the radiation envi-ronment in Ukraine. The background $^{137}$Cs contamination has increased to 4–20 kBq/m$^2$ throughout much of Ukraine. A wide variety of factors, from natural conditions to the way the accident unfolded, caused the distribution of contamination to be extremely complex and varied with respect to radionuclide form and content.

For nearly 30 years before the accident, environmental monitoring had been performed by various entities under the State Committee for Hydrol-ogy and Meteorology, the Ministry of Health, and the Ministry of Agricul-ture and Food Products. Following the accident, radiation monitoring was performed using ground vehicles, people on foot in the region, and aircraft. Unit 4 was monitored regarding the release of radionuclides. Also, the air-borne propagation of radionuclides throughout Ukraine as well as the scale of transboundary radionuclide transport was monitored. The radionuclide contamination in the environment and food products was monitored. In addition, personal dosimetric measurements were taken of individuals involved in accident cleanup operations (also known as the liquidators). The results of this monitoring have been used by various government agencies in making decisions aimed at minimizing the impact of the accident.

The environmental monitoring showed the $^{137}$Cs and $^{90}$Sr levels in the surface layers of soil were 0.8–2.2 kBq/m$^2$ (0.02–0.06 Ci/km$^2$) [United Nations 1982; Morozov et al. 1990; Nosko et al. 1994] before the accident. The origin of this activity is atmospheric testing of nuclear weapons from 1946—1963 (global fallout).

Virtually no systematic studies of plutonium and americium distribution in the environment were performed in Ukraine until 1986. Various papers [e.g., Eisenbud and Gesell 1991] by foreign and domestic researchers quote surface contamination values of 10–60 Bq/m$^2$ (0.3–1.7 mCi/km$^2$) for the plutonium and americium contamination at Northern Hemisphere latitudes corresponding to the latitude of Ukraine (44–52 degrees north latitude) before the accident. The gamma-ray radiation background 1 m above the surface ranged from 4–20 $\mu$R/h.

Until late 1987, most of the attention was focused on Kiev, Zhitomir, and Chernigov Oblasts, where contamination levels in excess of 5.5 x $10^7$ Bq/m$^2$ were observed, mostly in the West Plume. During this same year, locally produced food products with elevated levels of $^{137}$Cs were detected in the public food supply in Rivno and Volynsk Oblasts: milk with more than 370 Bq/L and mushrooms with 3,700 Bq/kg. Surveys have been performed in Southern Ukraine since 1988, and were later expanded to cover the entire country.

By late 1992, the entire country had been covered by an aerial gamma-spectrometric survey. The survey determined that several areas— with a total area of approximately 500 km$^2$—had $^{137}$Cs contamination levels greater than 37 kBq/m$^2$.

The aerial surveys enabled us to construct a map of the $^{137}$Cs contamination throughout the entire country.

In looking at cesium contamination in Ukraine, the West Plume (a pattern of radioactive contamination from the accident) had a significant impact. The radioactive contamination in the West Plume was basically deposited on 26–27 April 1986. Within Ukraine, this plume covers Kiev and Zhitomir Oblasts, as well as the northern portion of Rivno Oblast and the northeast portion of Volynsk Oblast. The area within the plume has an average contamination of 20–100 kBq/m$^2$, with individual spots having $^{137}$Cs contamination levels as high as 190 kBq/m$^2$. The Ukrainian Polessye, especially the portions located in Rivno and Volynsk Oblasts, have anomalously high soil-to-vegetation transfer coefficients, which has a substantial effect on the dose structure for the people living in these areas and consuming locally produced food products.

The South Plume passes through Kiev, Cherkassy, and Kirovograd Oblasts, as well as portions of Vinnitska, Odessa, and Nikolaevsk Oblasts and comes in next with respect to area and contamination level. In Kiev, Cherkassy, and the northwest portion of Kirovograd Oblast, this plume is wide, with contamination levels of up to 100 kBq/m$^2$, but with isolated areas having contamination levels as high as 185 kBq/km$^2$. In the Vinnitska, Odessa, and Nikolaevsk Oblasts, the contamination is spotty, with cesium contamination levels of 10–40 kBq/m$^2$ in most areas. In addition, the South Plume includes a spotty band of contamination that passes through portions of Vinnitsa, Khmel'nitskii, Ternopol', Ivano-Frankovsk, and Chernovits Oblasts. The predominant contamination levels range from 10–40 kBq/m$^2$, with isolated areas having contamination levels of 100–150 kBq/m$^2$.

Significant $^{137}$Cs contamination has also been observed in the western and northeastern portions of Chernigov Oblast (maximum value 185 kBq/m$^2$), northern Sumy, Donetsk, Lugansk, and Khar'kov Oblasts (generally with maximum values of 40 kBq/m$^2$).

Lower contamination levels were found in other parts of Ukraine. Certain areas in Western Ukraine (L'vov Oblast, part of Ternopol' Oblast, the Trans-Carpathian Oblast, and Khmel'nitskii Oblast, and adjacent portions of other oblasts), Southern Ukraine (Kherson Oblast, a portion of Zaporozh'e Oblast, Dnepropetrovsk Oblast, Nikolaev Oblast, Odessa Oblast, and the Crimea), as well as the northeast portions of Chernigov and Sumy Oblasts typically have $^{137}$Cs contamination levels below 4 kBq/m$^2$.

Ukraine amounts to almost 6% of Europe by area, while it received 14% of the total $^{137}$Cs fallout. Analysis of the spatial distribution of $^{137}$Cs on Ukrainian territory indicated that almost 40% of the total amount of $^{137}$Cs in Ukraine excluding buried radioactive waste in the Exclusion Zone is within 100 km of Chornobyl NPP, which is just slightly more than 3% of the total area of Ukraine (see Table 5.2.2).

**TABLE 5.2.2.** Distribution of $^{137}$Cs on Ukrainian Territory

| Inner Radius of Contamination Zone ($R_{in}$), km | Outer Radius of Contamination Zone ($R_{out}$), km | Area within the Zone | | $^{137}$Cs Activity Within a Specific Zone as a Percentage of the Entire Zone's Contamination (Q) | | $^{137}$Cs Contamination Level (q), kBq/m$^2$ |
|---|---|---|---|---|---|---|
| | | km$^2$ | % of Ukraine | Bq x 10$^{15}$ | % | |
| 0 | 10 | 314 | 0.05 | 1.58 | 11.81 | 5030 |
| 10 | 20 | 682 | 0.11 | 0.73 | 5.43 | 1065 |
| 20 | 30 | 771 | 0.13 | 0.27 | 1.99 | 340 |
| 0 | 30 | 1,767 | 0.29 | 2.57 | 19.23 | 1455 |
| 30 | 40 | 1,198 | 0.20 | 0.36 | 2.66 | 297 |
| 40 | 100 | 17,744 | 2.94 | 2.42 | 18.09 | 136 |
| 0 | 100 | 20,709 | 3.43 | 5.35 | 39.98 | 258 |
| 100 | 200 | 59,532 | 9.86 | 2.03 | 15.17 | 34 |
| 200 | 400 | 209,143 | 34.64 | 3.29 | 24.62 | 16 |
| 400 | 600 | 205,587 | 34.06 | 1.63 | 12.19 | 8 |
| 600 | 830 | 108,729 | 18.01 | 1.07 | 8.04 | 10 |
| 0 | 830 | 603,700 | 100 | 13.38 | 100 | 22 |

Elevated concentrations of transuranic elements and nuclear-fuel fission and decay products from the Chornobyl NPP accident are found throughout Ukrainian environment. Every oblast has large radionuclide fallout fields at concentrations four to five times higher than the pre-accident background. Natural and anthropogenic variations in landscape and surface roughness, as well as microclimatic conditions, all contributed to the "spotty" structure of the contamination.

More than 13,000 km$^2$ of Ukraine has $^{137}$Cs contamination levels greater than 100 kBq/m$^2$ and nearly half of this area is forested. As of 1 July 1997, more than 43,500 km$^2$ of area in Ukraine has contamination exceeding 40 kBq/m$^2$ not including the city of Kiev. In this 43,500 km$^2$ area, 30% was observed to have isolated local anomalies with contamination levels greater than 40 kBq/m$^2$.

## 5.3  SURFACE WATER CONTAMINATION

Surface water bodies can be found near the Chornobyl NPP. These bodies of water include the Dnepr River, the Pripyat River, and the Kiev Reservoir. The Dnepr River rises southwest of Moscow and flows in a generally southern direction to the Black Sea, passing through Kiev on the way. This river—which is 2,290 km long—is a long series of reservoirs and the major waterway in Ukraine. The Pripyat River is a tributary of the Dnepr River. The Pripyat River rises in northwestern Ukraine near the Polish border and flows eastward through Ukraine and then Belarus through a flat, forested, and swampy basin known as the Pripyat Marshes to the Dnepr River. The Dnepr and Pripyat Rivers feed into the Kiev Reservoir, which is located just north of the Ukrainian capital city, Kiev. The Kiev Reservoir drains through a series of reservoirs to the Black Sea.

### 5.3.1  Early Stages of Surface Water Contamination

A body of water can become contaminated through radioactive atmospheric fallout and contact between contaminated air and the water surface. A variety of secondary contamination paths also exist. These paths include runoff, flow of contaminated water from more contaminated bodies of water into less contaminated bodies of water, mass exchange between bottom sediments and bodies of water, and discharge of contaminated subsurface water into the surfaces of bodies of water.

A wide range of scientific reports and literature is available on radioactive contamination in rivers, lakes, and reservoirs [Voitsekhovich 1992; Ministry for Chornobyl Affairs 1996; Izrael et al. 1987; Vakulovskii et al. 1989]. The surface water contamination shortly after the accident (in May 1986) included several dozen radionuclides, with the following radionuclides making the largest contributions to the ecosystem dose rate: $^{141}$Ce, $^{144}$Ce, $^{103}$Ru, $^{140}$Ba, $^{131}$I, $^{95}$Zr, $^{95}$Nb, $^{140}$La, $^{134}$Cs, and $^{137}$Cs. Other uranium fission products were detectable but were present only in trace quantities. Each radionuclide made a different contribution to the total activity in the water. Since 1987, however, radionuclides such as $^{137}$Cs and $^{90}$Sr have been the main contributors to surface water contamination.

The highest contamination levels were recorded during the first 10 days of May 1986, when the fallout rate was highest; once the fallout stopped, the waterborne contamination level began to drop. The total waterborne activity in the Pripyat River decreased from approximately $10^6$ Bq/L during the first few days following the accident to $10^3$–$10^4$ Bq/L by early June. The highest contamination level for $^{90}$Sr in the Pripyat River was 20 Bq/L (data provided by NPO Taifun), with up to 100 Bq/L (or even higher) being recorded in bodies of water in the inner Exclusion Zone. The inner zone is defined as a circle with a 10-km radius with the Chornobyl NPP at the center. The highest $^{137}$Cs contamination levels in Pripyat River water (on the order of $10^3$ Bq/L) were also observed in early May 1986, when $^{239}$Pu concentrations of approximately 1 Bq/L were observed; by August, these concentrations had decreased by an order of magnitude. Following the accident, a sharp increase in $^{131}$I was observed in the Dnepr River, which was regularly monitored for $^{131}$I at Kiev by the Kiev health authorities and Taifun [Vakulovsky et al. 1994].

Radionuclides from Chornobyl NPP accident were clearly detectable in many rivers throughout the European sector of the former Soviet Union and Western Europe. Many of these rivers returned to the pre-accident background radioactive levels quite slowly, and some have still not returned to the pre-accident levels 12 years after the accident. Despite the current low radionuclide levels in these rivers, elevated levels of $^{137}$Cs and $^{90}$Sr are observed on an annual basis during the spring runoff and rainy season.

Table 5.3.1 provides data on the contamination levels for Ukrainian rivers in the near and far fallout zones.

### 5.3.2  Bodies of Water in the Exclusion Zone

5.3.2.1  **Open and Closed Oxbows Along the Pripyat River.** A significant fraction of the Exclusion Zone lies within the Pripyat or Dnepr floodplain or the interfluve between the Pripyat and Dnepr. These areas contain a large number of bodies of water that take the form of open oxbows (inlets), closed oxbows (detached old riverbeds), or lakes resulting from the behavior of the river channel (Figure 5.3.1). Lake Azbuchin (the section of the floodplain adjacent to Chornobyl NPP on the right bank of the river), Lake

**TABLE 5.3.1.** Concentrations of Major Radionuclides in Ukrainian Rivers After the Accident (1986)

| River | Sample Collection Date (1986) | $^{131}I$ (Bq/L) | $^{141}Ce$ (Bq/L) | $^{103}Ru$ (Bq/L) | $^{95}Zr$ (Bq/L) | $^{140}Ba$ (Bq/L) | $^{137}Cs$ (Bq/L) | $^{90}Sr$ (Bq/L) |
|---|---|---|---|---|---|---|---|---|
| Pripyat River | 1 May | 2,109 | 407 | 555 | 407 | 1,406 | 247.9 | 10 |
| | 2 May | 444 | — | 814 | 1,554 | 2,220 | 555 | |
| | 6 May | 814 | 90 | 170 | 166.5 | 166.5 | 1,591 | |
| Teterev River | 3 May | 1,998 | 666 | 703 | 1,443 | 1,258 | | |
| Dnepr River (Kiev) | 3 May | 1,295 | 333 | — | 703 | 703 | 1,295 | — |

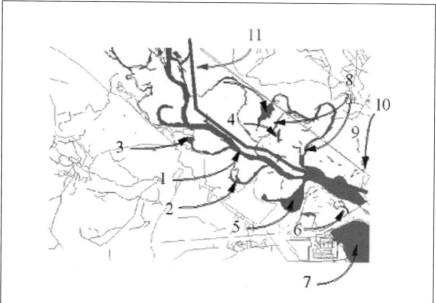

1. Pripyat River channel, dredged between 1993 and 1997
2. Creek Semikhody, cut off in 1986
3. Creek Shepelinskii Starik, cut off in 1986
4. Lake Glubokoe and Lake Vershina
5. Creek Pripyati (Yanovskii Creek), cut off in 1986
6. Lake Azbuchin (partially covered by river sand 1997–1998)
7. Cooling pond
8. Krasnyanskii Starik and drainage canal
9. Reclaimed area (drainage system) and control facilities
10. Control facilities for water from reclaimed area
11. Dike with drainage channel, left bank (1993)

**FIGURE 5.3.1.** Bodies of water and hydraulic structures in the inner Exclusion Zone around Chornobyl NPP

Glubokoe, and Lake Vershina are overlapped by the Krasnyanskii Starik oxbow lake in the floodplain on the left bank of the river. Following completion of the reservoir dam in 1993, these lakes turned out to be within the dammed area on the left-bank floodplain [Remis 1996].

Most of the open and closed oxbows have a continuous or seasonal hydraulic connection with the main river channel. The high groundwater level (0.5–2 m below the surface) near most of the oxbows, coupled with the highly permeable alluvial sand deposits in the floodplain, also facilitates partial water exchange between the oxbows and the main channel.

The bodies of water and river surfaces were initially contaminated by atmospheric fallout similar in composition to the fuel matrix (in the inner zone adjacent to Chornobyl NPP) [Makhon'ko 1990; Voitsekhovich 1995]. Shortly after the accident, the contamination levels in the rivers and bodies of water fluctuated as atmospheric fallout continued and as radionuclide migration occurred in the environment. The contamination levels resulting from this fallout have stabilized over the years since the accident.

Contamination levels have decreased in the rivers as a result of water exchange and the scouring of the river bottoms. The river bottoms have been scoured each year during floods (both the annual spring floods and others). So, the radioactive material in the bottom sediments no longer plays an important role as a secondary source of contamination for river water.

However, enclosed bodies of water in the Exclusion Zone show much higher levels of radioactive contamination. This higher contamination is because of the reduced water exchange and the increased role played by sedimentary deposition of radionuclides. The bottom deposits in enclosed bodies of water have much higher levels of radioactive contamination associated with the initial fallout in 1986 than the river-channel and elutionary soil in the riverbed.

Thus, the current radioactive contamination level in most enclosed bodies of water will be determined by the rate of mobile-radionuclide mass exchange between the contaminated bottom deposits and the water, as well as radioactive runoff from adjacent areas. The bottom deposits in some lakes (e.g., Lake Azbuchin, Lake Glubokoe, Creek Pripyati, the Chornobyl NPP cooling pond) have such high contamination levels that they could be classified as low- or medium-level radioactive waste for disposition purposes.

Table 5.3.2 lists the typical mean annual contamination levels for the most important bodies of water in the Exclusion Zone as derived from composite estimates obtained during detailed surveys performed by Ukrainian Scientific Research Institute for Hydrology and Meteorology (UkrNIGMI) in 1990–1991. Table 5.3.3 lists typical mean annual contamination data obtained between 1987 and 1995 for the most important Exclusion Zone

**TABLE 5.3.2.** Typical Contamination Levels for Floodplain Lakes in the Exclusion Zone (1990 - 1991)

| Body of Water | Surface Area km² | Volume of Water 10⁻⁴ km³ | Radionuclide Content of Water, Bq/L $^{137}$Cs | $^{90}$Sr | Bottom Contamination, kBq/m² $^{137}$Cs | $^{90}$Sr |
|---|---|---|---|---|---|---|
| Rukav Murovka | 0.23 | 5.75 | 0.4–0.7 | 1.5–3.7 | 296 | 185 |
| Krasnyanskii Starik | 0.32 | 0.1 | 7.4–18.5 | 37–92.5 | 7,400 | 5,920 |
| Lake Glubokoe | 0.10 | 5.0 | 37–74 | 200–400 | 29,600 | 24,420 |
| Lake Vershina | 0.03 | 0.8 | 29.6–37 | 200–300 | 25,160 | 18,500 |
| Oxbow (1) | 0.08 | 1.9 | 29.6–37 | 200–300 | 6,660 | 4,810 |
| Oxbow (2) | 0.043 | 12.5 | 29.6–37 | 300–400 | 22,200 | 19,980 |
| Oxbow (3) | 0.024 | 36.0 | 29.6–37 | 300–400 | 9,250 | 8,140 |
| Lake Beloe | 0.021 | 63.0 | 5.6–7.4 | 80–100 | 7,400 | 4,440 |
| Lake Azbuchin | 0.05 | 1.5 | 11.1–18.5 | 100–200 | 12,950 | 9,620 |
| Creek Pripyati | 0.82 | 33.0 | 7.4–14.8 | 80–100 | 13,320 | 8,140 |
| Creek Semikhody | 0.20 | 5.0 | 3.7–7.4 | 29.6–37 | 8,140 | 7,400 |
| Creek Novoshepelichi | 0.21 | 4.5 | 2.2–5.6 | 3.7–7.4 | 2,220 | 1,850 |

**TABLE 5.3.3.** Mean Annual Concentrations of Radionuclides in Several Inner Exclusion Zone Water Bodies from 1987 to 1995

| Bq/L | 1987 | 1988 | 1989 | 1990 | 1991 | 1992 | 1993 | 1994 | 1995 | 1996 | 1997 |
|---|---|---|---|---|---|---|---|---|---|---|---|
| | | | | | *Cooling Pond* | | | | | | |
| $^{137}$Cs | 70.30 | 32.19 | 21.09 | 12.21 | 8.14 | 5.18 | 6.29 | 5.55 | 5.18 | 4.15 | 2,75 |
| $^{90}$Sr | 7.4 | 15.91 | 15.54 | 8.51 | 7.03 | 4.81 | 4.44 | 3.37 | 3.29 | 2.75 | 2.3 |
| | | | | | *Glinitsa River* | | | | | | |
| $^{137}$Cs | 3.7 | 1.48 | 4.8 | 0.37 | 0.43 | 0.3 | 0.48 | 0.41 | 0.37 | 0.37 | 0,32 |
| $^{90}$Sr | 5.18 | 5.18 | 8.14 | 5.55 | 7.77 | 3.11 | 7.77 | 8.14 | 7.47 | 5.95 | 4.9 |
| | | | | | *Creek Pripyati* | | | | | | |
| $^{137}$Cs | 25.16 | 9.62 | 11.1 | 12.58 | 7.03 | 11.47 | 9.99 | 8.51 | 6.14 | 6.9 | 5,1 |
| $^{90}$Sr | 133.20 | 88 | 13.69 | 104.71 | 74 | 81.4 | 81.4 | 78.81 | 61.49 | 44 | 29.7 |
| | | | | | *Creek Semikhody* | | | | | | |
| $^{137}$Cs | 18.87 | 7.4 | 6.66 | 5.18 | 4.44 | 3.7 | 3.59 | 3.15 | 1.89 | 1.55 | 1,67 |
| $^{90}$Sr | — | 37 | 40.7 | 48.1 | 70.3 | 65.49 | 55.5 | 62.9 | 42.55 | 28.4 | 18.2 |

bodies of water [Ogorodnikov and Kanivets 1993; Voitsekhovitch et al. 1996a].

Creek Pripyati (Yanovskii Creek) and Glubokoe Lake are the most extensively studied bodies of water with respect to radioactive contamination. These are also the most contaminated bodies of water in the northwestern sector and are located 1.5–5 km from the destroyed reactor. During construction of Chornobyl NPP and the city of Pripyat, Creek Pripyati was enlarged and dredged to accommodate a river port. The sand dredged from Creek Pripyati was deposited on the northwest bank and used as a sand foundation for construction of a new district for the city of Pripyat (currently the site of the Peschanoe Plato Interim Radioactive Waste Storage Site).

By fall 1986, bulk stone and sandbags had been deposited into Creek Pripyati, along with other bodies of water near the Chornobyl NPP sanitary exclusion zone, to stop the contaminated surface water from flowing out of these bodies and into other waterways. The stone and sandbags provide a flow restriction sometimes called a weep or filtration prism.

Various organizations have collected data on the contamination of Creek Pripyati. The available observational data on the radionuclide concentration in the water and bottom sediment vary widely. This is because of imperfections and inconsistencies in the sample collection and radiation monitoring methods used by various organizations. The data also varies because of the partial flushing by river water during rain-driven flooding in the summer of 1988, high water in 1987 and 1994, and flooding in the winters of 1991 and 1994 as a result of ice jams on the Pripyat River. After a flood, the high water on the river recedes, and the water level is higher in Creek Pripyati than in the main river channel. This high water level ensures that there is a virtually constant flow of contaminated water (with a flow rate ranging from 50–200 L/s) from Creek Pripyati into the river through the filtration prism. Bottom sediments and discharge of radioactive groundwater are the main sources of contamination for the water in Creek Pripyati. Data obtained by the Dosimetric Monitoring Administration in March 1995 [Ogorodnikov and Kanivets 1993] suggest that the Creek Pripyati bottom sediment contains $7.4 \times 10^{12}$–$8.1 \times 10^{12}$ Bq (200–220 Ci) of $^{137}$Cs and approximately $8.51 \times 10^{12}$ Bq (230 Ci) of $^{90}$Sr (previous estimates were 150–180 Ci) and up to $3.7 \times 10^{11}$ Bq (10 Ci) of plutonium isotopes and $^{241}$Am. More than 90% of the estimated total radioactivity in Creek Pripyati is concentrated at depths of 6–13 m, where regions of stable silt accumulation have been identified.

5.3.2.2 **Cooling Pond.** Over the years since the accident, many researchers have performed detailed studies of the radioactive contamination in the Chornobyl NPP cooling pond [Ogorodnikov and Kanivets 1993; Ukrainian-Canadian Expedition 1995; Ukrainian Unit 1995]. The available data suggest that up to 7.4 x $10^{15}$ Bq (200,000 Ci) of various fission products were initially deposited in the cooling pond in the form of fallout. However, the water rapidly became cleaner through radionuclide decay and a variety of physical, chemical, and biological processes. By 1991, the total amount of $^{137}$Cs in the cooling pond water and bottom sediment was estimated to be 1.3 x $10^{14}$–1.7 x $10^{14}$ Bq (3,500–4,500 Ci). UkrNIGMI has estimated the total amount of $^{90}$Sr in the cooling pond bottom sediment and water at 5.55 x $10^{13}$–7.4 x $10^{13}$ Bq (1,500–2,000 Ci) in 1991. Various estimates of the transuranic elements present at the bottom of the cooling pond range from 5.55 x $10^{12}$–7.4 x $10^{12}$ Bq (150–200 Ci); however, the filtrational outflow causes virtually no migration of the transuranic elements beyond the bottom of the pond. The horizontal and vertical structure of the contamination levels in the bottom sediment is highly nonuniform; this will undoubtedly make it more difficult to decontaminate the bottom sediments should the decision be made to do so in the future.

In 1991–1992, the pond water contained a total of 7.4 x $10^{11}$–1.1 x $10^{12}$ Bq (20–30 Ci) of $^{90}$Sr; however, in 1994–1995, it contained at most 5.55 x $10^{11}$ Bq (15 Ci). The annual filtrational water loss is approximately equal to the total pond capacity at the normal support level (140–145 million m$^3$), with the 1994–1995 observational data showing no substantial adsorption of $^{90}$Sr on soil in the filtration zone (with the process that has developed over time in the equilibrium phase).

Since 1992, the level of radioactive contamination in the pond water has exhibited a relatively stable decreasing trend. Under these conditions, the process used to control the water level in the pond becomes very important. Decreasing the water level in the pond will reduce the head gradient and, thus, the filtration rate. If we assume that the observed trend lines for the rate of change in the radionuclide content of the cooling pond are identical to those for 1993–1995, and cut the relative head in half by lowering the water level (which can be done once Chornobyl NPP is shut down) even an approximate calculation suggests that the maximum total annual infiltration into the Pripyat River from the cooling pond during the next 3–5 years will be 2 x $10^{12}$–4 x $10^{12}$ Bq/yr (8–10 Ci/yr), and should be even lower.

The infiltration from the Chornobyl NPP cooling pond into the Pripyat River is partially discharged into the river via the watersheds of 11 intermittent streams that originate at the northern drainage canal, as well as the southern drainage canal, which discharges into the Glinitsa River. The intermittent filtration streams have a total mean monthly flow rate that ranges from 1–1.5 m$^3$/s, depending on season, with mean contamination levels (averaged on an annual basis from 1994–1996) ranging from 0.6–0.9 kBq/L and 185–296 Bq/L for $^{90}$Sr and $^{137}$Cs, respectively. Filtration stream PK-17 is an exception; this stream may be as much as 5–10 times more contaminated than the other streams. Stream PK-17 is more contaminated because it passes through several highly contaminated bodies of water in the floodplain between the cooling pond levee and the Pripyat River. The contamination levels in the filtration streams have decreased over the past 5 years, consistent with the general trend of reduced contamination in the Chornobyl NPP cooling pond. Although the total annual $^{90}$Sr discharge into the filtration water from the cooling pond via the north drainage filtration streams was minimal in 1997, from 1992 to 1996, it was 2 x $10^{11}$–4 x $10^{11}$ Bq/yr (6–12 Ci/yr).

### 5.3.3 Minor Rivers and Intermittent Streams in the Exclusion Zone

The hydrographic network in the Exclusion Zone is relatively highly developed. The drainage network has a mean density of 0.3–0.4 km/km². All of the minor rivers whose drainage networks lie to some extent within the Exclusion Zone, with the exception of the Braginka River, are first-order, or perhaps even second- or third-order, tributaries of the Pripyat River, with the higher-order tributaries flowing into the watersheds of the Uzh, Grezlya, Il'ya, Bober, or Veresnya Rivers. Several researchers have analyzed the contamination of minor rivers in the Exclusion Zone [Remis 1996; Borsilov 1996, European Union 1996; Ministry for Chornobyl Affairs 1996]. The radioactive contamination of minor rivers in the Exclusion Zone is associated with a variety of different sources, currently radioactive runoff from the adjacent watersheds, and geomorphological landscape forms, to the extent that the latter impact runoff conditions. The specific runoff coefficients for $^{137}$Cs are decreasing with time, due to radionuclide fixation in the watershed soil and vertical-migration-induced depletion of the radionuclide content in the soil surface layer. The runoff rate for $^{90}$Sr is also gradually decreasing but in this case largely due to a substantial increase in the depth of maximum contamination (i.e., the contamination is penetrating further into the soil).

The main tributary of the Pripyat River in the Exclusion Zone is the Uzh River, whose watershed covers the southern edge of the Exclusion Zone, where the $^{90}$Sr contamination levels are relatively low. Virtually the entire Uzh watershed, from the city of Pripyat to the urban settlement of Polessko, lies within the Western Cesium Plume (also known as the Polessko-Narodichskii Plume). Up to 80% of the river's annual flow of 20–150 m³/s is produced during spring high water. The river is subject to flash flooding, with brief peak flow rates of 100 m³/s or higher. Over the years since the accident, the observed concentration of $^{90}$Sr in the river at Cherevach was 185–1,850 Bq/L (5 x $10^{-9}$–50 x $10^{-9}$ Ci/L). From 1991–1996, the maximum observed concentrations of radioactive cesium was 370–555 Bq/L (10 x $10^{-9}$–15 x $10^{-9}$ Ci/L), with mean values of 110–220 Bq/L (3 x $10^{-9}$ –6 x $10^{-9}$ Ci/L). The annual discharge of $^{90}$Sr into the waters of Uzh and Pripyat Rivers during low-water years such as 1992 and 1995 is 1.5 x $10^{11}$–1.9 x $10^{11}$ Bq (4–5 Ci). During years with a typical water level, the annual discharge of $^{90}$Sr is 5.5 x $10^{11}$–7 x $10^{11}$ Bq (15–19 Ci); this amounts to 2–5% of the total $^{90}$Sr radioactivity coming into the Kiev Reservoir.

### 5.3.4 **Radioactive Contamination in the Pripyat and Dnepr Rivers**

Since the accident, the radioactive contamination in the Pripyat and Dnepr Rivers as a function of time was completely determined by the hydrological runoff regime in the watershed, the flow rate in the river channel, and the transformations between various physical and chemical forms of $^{90}$Sr and $^{137}$Cs in the soil of the Pripyat River watershed and floodplain. The basic observations of water flow and radionuclides from the watersheds of the Dnepr and Pripyat were performed at the Nedanchichi and Chornobyl monitoring stations, respectively. The transport balance for radioactive substances in the Dnepr River watershed and reservoirs along the Dnepr River was determined by analyzing all available monitoring observations. The main contribution to the radioactive runoff into the Dnepr River in 1986 came from the Sozh River, which received most of its water from highly contaminated areas in Gomel' Oblast, Belarus, as well as the Bryansk-Tula spot of cesium fallout. In 1993, the water contamination level at the mouth of the Dnepr River began to stabilize; by 1995, the contamination levels approached the pre-accident levels for a large fraction of the year.

The Pripyat River is contaminated by radioactive runoff from the watersheds and floodplains of its tributaries. Figure 5.3.2a shows the river contamination as a function of time starting in May 1986 through May 1995. The data indicate that before full implementation of water management measures (e.g., sandbags) on the left-bank floodplain in 1993, virtually every flood related to spring runoffs or ice jams led to a factor of 5–10 increase in the downstream concentration of $^{90}$Sr.

Following full implementation of water management measures along the left bank, drainage of radioactivity into the Pripyat River was substantially reduced, and runoff from the right-bank floodplain and inner 5- to 10-km Exclusion Zone became the main sources of radioactivity.

The monthly average variations in radioactive contamination for the Dnepr River (Figure 5.3.2b) are not as great as those for the Pripyat River, and principally reflect seasonal variations in runoff. The seasonal variations in the $^{90}$Sr and $^{137}$Cs concentrations in water from the Dnepr River are observed to be correlated, indicating that the distribution of contamination throughout the watershed is relatively uniform and that the amount of radioactive runoff is determined by the catchment area within the watershed.

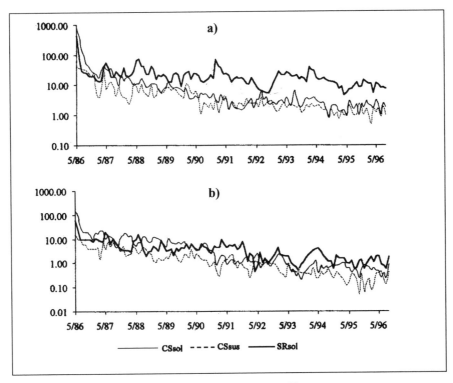

**FIGURE 5.3.2.** Ten-day mean values for [137]Cs and [90]Sr radionuclide concentrations (10[-9] Ci/L) for 1986–1995 (a) in water from the Pripyat River at Chornobyl, and (b) the Dnepr River at Nedanchichi (b). Cssol-soluble [137]Cs, CSsus-suspended [137]Cs, SRsol-soluble [90]Sr.

Moreover, radioactive contamination is steadily decreasing with time in both the Pripyat and Dnepr Rivers (Table 5.3.4).

The waterborne contamination in the Pripyat River during various seasons of the year largely depends on the local runoff conditions for radioactive materials in the Exclusion Zone. The [90]Sr runoff from the Exclusion Zone is largely determined by localized washouts and runoff over the floodplain (which may occur independently of the seasonal water flow variations, but can still lead to flooding of contaminated land during ice jams), as well as control operations involving contaminated filtration or drainage water. Thus, all of the significant peaks in the [90]Sr content of the water shown in Figure 5.3.2 (1988, 1991, 1994, etc.) involved water moving over the floodplain in the inner Exclusion Zone.

After full implementation of water management measures, a significant amount of radioactive material continued to be removed from the floodplain and into the rivers. This was because flooding of floodplain soil on the right

**TABLE 5.3.4.** Radioactive Runoff from the Pripyat and Dnepr Rivers into the Kiev Reservoir Following the Chornobyl Accident, Bq

| | Pripyat River (at Chornobyl) | | | Dnepr River (at Nedanchichi) | | |
|---|---|---|---|---|---|---|
| | $^{137}Cs$ | | $^{90}Sr$ | $^{137}Cs$ | | $^{90}Sr$ |
| Year | soluble | suspended | soluble | soluble | suspended | soluble |
| 1986* | $5.64 \times 10^{13}$ | $9.7 \times 10^{12}$ | $2.76 \times 10^{13}$ | $2.73 \times 10^{13}$ | $3.63 \times 10^{12}$ | $1.07 \times 10^{13}$ |
| 1987 | $9.03 \times 10^{12}$ | $3.7 \times 10^{12}$ | $1.04 \times 10^{13}$ | $9.66 \times 10^{12}$ | $4.18 \times 10^{12}$ | $7.99 \times 10^{12}$ |
| 1988 | $5.92 \times 10^{12}$ | $3.6 \times 10^{12}$ | $1.87 \times 10^{13}$ | $6.73 \times 10^{12}$ | $2.55 \times 10^{12}$ | $5.18 \times 10^{12}$ |
| 1989 | $3.63 \times 10^{12}$ | $2.8 \times 10^{12}$ | $8.92 \times 10^{12}$ | $5.8 \times 10^{12}$ | $1.26 \times 10^{12}$ | $3.59 \times 10^{12}$ |
| 1990 | $2.29 \times 10^{12}$ | $2.3 \times 10^{12}$ | $7.88 \times 10^{12}$ | $3.7 \times 10^{12}$ | $1.44 \times 10^{12}$ | $3.7 \times 10^{12}$ |
| 1991 | $1.44 \times 10^{12}$ | $1.4 \times 10^{12}$ | $1.44 \times 10^{13}$ | $1.07 \times 10^{12}$ | $9.62 \times 10^{11}$ | $4.51 \times 10^{12}$ |
| 1992 | $9.25 \times 10^{11}$ | $8.1 \times 10^{11}$ | $4.14 \times 10^{12}$ | $8.14 \times 10^{11}$ | $4.81 \times 10^{11}$ | $6.66 \times 10^{11}$ |
| 1993 | $2.18 \times 10^{12}$ | $1.3 \times 10^{12}$ | $1.48 \times 10^{13}$ | $6.29 \times 10^{11}$ | $3.33 \times 10^{11}$ | $1.18 \times 10^{12}$ |
| 1994 | $1.78 \times 10^{12}$ | $1.1 \times 10^{12}$ | $1.31 \times 10^{13}$ | $7.4 \times 10^{11}$ | $2.96 \times 10^{11}$ | $2.81 \times 10^{12}$ |
| 1995 | $7.4 \times 10^{11}$ | $5.6 \times 10^{11}$ | $3.63 \times 10^{12}$ | $5.18 \times 10^{11}$ | $2.22 \times 10^{11}$ | $8.14 \times 10^{11}$ |
| 1996 | $8.5 \times 10^{11}$ | $4.8 \times 10^{11}$ | $3.48 \times 10^{12}$ | $1.85 \times 10^{11}$ | $1.11 \times 10^{11}$ | $7.03 \times 10^{11}$ |
| 1997 | $8.9 \times 10^{11}$ | $6.3 \times 10^{11}$ | $2.59 \times 10^{12}$ | $1.85 \times 10^{11}$ | $7.4 \times 10^{10}$ | $6.66 \times 10^{11}$ |

* Estimate

bank of the Exclusion Zone between the village of Benevka and the cooling pond levee caused radionuclides (primarily $^{90}Sr$) to be washed into the main watercourse.

The trends for conversion of $^{137}Cs$ and $^{90}Sr$ into other physical and chemical forms with time indicate that $^{137}Cs$ fixation in Exclusion Zone soil was much more rapid than $^{90}Sr$. As a result, $^{137}Cs$ is no longer removed by flash floods, snowmelt, or flooding within the floodplain. This has been confirmed through observations of the radionuclide balance in the Exclusion Zone (Figure 5.3.3). For more information on the physical and chemical transformations, see Section 5.5.

As is evident in from Figure 5.3.3, the Exclusion Zone was not a significant source of $^{137}Cs$ during any of the Pripyat River water flow scenarios occurring between 1993 and 1996. In contrast to $^{90}Sr$, most of the $^{137}Cs$ originates from areas outside the Exclusion Zone.

During the high-water periods between 1993 and 1996, up to 70–80% of the $^{137}Cs$ in the Pripyat River flow was generated outside the Exclusion Zone in the watersheds of Ukrainian Polessye and Belarus. The Exclusion Zone, radioactively contaminated bodies of water serving as sources of runoff, and radioactively contaminated bodies of water subject to flushing during periods of high water flow were virtually the only significant sources of Chornobyl-generated $^{90}Sr$ to the Pripyat River. During periods of high water

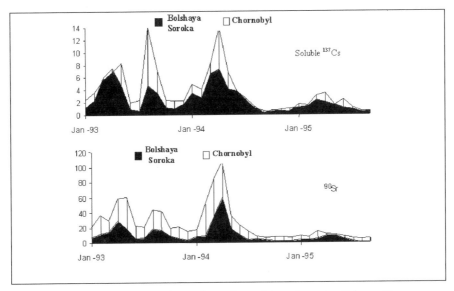

**FIGURE 5.3.3.** Transport balance for radionuclides along the Pripyat River (x 10$^{-9}$ Ci/L) at Bolshaya Soroka (where the river crosses the boundary of the Exclusion Zone on the Belorussian side) and Chornobyl (below the area which contributed the bulk of the radioactive runoff in 1993–1995)

flow, $^{90}$Sr-contaminated runoff into reservoirs along the Dnepr has been observed in all of the reservoirs along the Dnepr River.

### 5.3.5 Radioactive Contamination of Dnepr River Reservoirs

**5.3.5.1 Radioactive Contamination of Water During Initial Fallout.** Dense radioactive fallout south of Chornobyl NPP began on 29 April [Izrael 1984]. Kiev and Kanev Reservoirs were the reservoirs most affected by radioactive contamination from atmospheric fallout. The fallout continued through virtually the entire month of May, with the densest fallout being observed between 1 and 3 May.

The $^{90}$Sr and $^{137}$Cs fallout values for the Kiev and Kanev Reservoirs in Table 5.3.5 were calculated using contour maps showing the density of aerosol fallout for these radionuclides over the surface of each reservoir. The analysis was based on Ukrainian State Committee for Hydrology and Meteorology maps of the contamination in areas adjacent to the reservoirs. The fallout densities for the surfaces of reservoirs located downstream of the Dnepr River were determined by analyzing shore samples collected by UkrNIGMI between 1988 and 1991; global pre-Chornobyl soil contamination was taken into account [Makhon'ko et al. 1993; Moiseev and Ramzaev 1975]. No data were provided for the $^{90}$Sr aerosol fallout on the Kremenchug Reservoir or reservoirs downstream of the reservoir, as the measured results from the shore samples did not exceed the pre-Chornobyl global fallout values.

The data in Table 5.3.5 describe the $^{137}$Cs and $^{90}$Sr content of the reservoirs along the Dnepr River as of 1986. Over the subsequent years, the radionuclides in the Dnepr aqueous environment have been redistributed by hydrodynamic processes, processes internal to each body of water, and physical and chemical transformations.

**TABLE 5.3.5.**   Mean Area-Averaged Values of $^{137}$Cs and $^{90}$Sr Aerosol Fallout Following Chornobyl Accident

| Reservoir | $^{137}$Cs | | $^{90}$Sr | |
|---|---|---|---|---|
| | Bq | Bq/m$^2$ | Bq | Bq/m$^2$ |
| Kiev | 6.18 x 10$^{13}$ | 66,970 | 1.85 x 10$^{13}$ | 19,980 |
| Kanev | 9.44 x 10$^{12}$ | 14,060 | 3.33 x 10$^{12}$ | 4,810 |
| Kremenchug | 8.14 x 10$^{12}$ | 3,626 | | |
| Dneprodzerzhinsk | 1.3 x 10$^{12}$ | 2,294 | | |
| Dnepr | 3.7 x 10$^{11}$ | 888 | | |
| Kakhov | 1.3 x 10$^{12}$ | 592 | | |
| Total: | 8.23 x 10$^{13}$ | | 2.18 x 10$^{13}$* | |

*Values provided only for Kiev and Kanev Reservoirs.

5.3.5.2 **Radionuclide Inflow in River Water.** Since the initial fallout, the radiation in the Dnepr aqueous environment has been determined by the inflow of radionuclides from contaminated areas into the Dnepr River. The inflow from the Pripyat and Upper Dnepr Rivers has been and continues to be the main source of radionuclides for the Dnepr reservoirs. The radionuclide input from these rivers, as well as the Desna River, has a substantial impact on the contamination in all of the Dnepr reservoirs. The radionuclide input from all other rivers is insignificant by comparison. Despite the fairly significant fraction of the $^{137}$Cs and $^{90}$Sr that is provided by the Upper Dnepr River, it is still the most significant diluent for the radioactive runoff from the Pripyat River and other rivers. The Desna River significantly dilutes the contaminated water discharged through the Kiev Hydroelectric Power Plant into the mid-Dnepr reservoirs.

Estimates indicate that complete mixing of Pripyat River water with Dnepr River water in the "mixing bowl" formed by the Kiev Reservoir reduces the $^{137}$Cs concentration by a factor of 1.5 and the $^{90}$Sr concentration by as much as a factor of 2 (averaged over the 10-year observing period). Similarly, the estimated dilution factor for the Dnepr River by less contaminated water is a factor of 1.24 for $^{137}$Cs and 1.23 for $^{90}$Sr. These estimated dilution factors show only insignificant variation for specific years of observation since the accident, which suggests that the mean values can be used as rough predictive estimates for the radionuclide conversion coefficients on the lower Dnepr River.

Over the years since the accident, the radioactive contamination in the rivers, which has been determined by the radionuclide runoff conditions within the Chornobyl NPP watershed, floodplain, and site, has also determined the seasonal variations in reservoir contamination.

The main trend for the inflow of contaminated water has been for the total radionuclide runoff into the Dnepr reservoirs to decrease with time. From 1987–1997, the $^{137}$Cs runoff in the Pripyat River decreased by a factor of 5, while the $^{90}$Sr runoff decreased only slightly. Correspondingly, the $^{90}$Sr:$^{137}$Cs ratio in the runoff has increased over the years since the accident (in 1987, the ratio was 0.8; in 1997, it was 4).

Note that all the extreme increases in Pripyat River $^{90}$Sr concentration since the accident (1988, 1991, 1993, and 1994) were related to floodplain flooding in the inner Exclusion Zone. These floods later affected the radiation environment in all of the Dnepr River reservoirs, giving rise to a wave of elevated $^{90}$Sr concentrations that could be traced all the way to the Kakhovka Reservoir and the Black Sea (Figure 5.3.4).

The general trend for the radionuclide runoff to decrease with time becomes especially apparent by comparing the radiation environments in the rivers for 1992 and 1995, which turn out to have had similarly low runoff levels and similar runoff formation conditions. This trend is due to two factors acting in concert: 1) fixation of radionuclides in soil particles throughout the watersheds, and 2) a decrease in the exchangeable forms of radionuclides as a result of leaching into the drainage network, of infiltration, and of downward penetration of radionuclides into the soil.

The reduction in $^{137}$Cs transport was independent of variations in water flow. The fraction of $^{137}$Cs transported in suspended form increased from 29% in 1987 to 47% in 1991. Although monitoring data indicate that this ratio has stabilized in recent years, it seems more likely for $^{137}$Cs transport into the drainage network during floods to involve the suspended form than the soluble form. Episodic observations performed on the Pripyat and Dnepr Rivers by UkrNIGMI from 1994 to 1996 invariably show that most of the $^{137}$Cs (50–60%) was transported in suspended form.

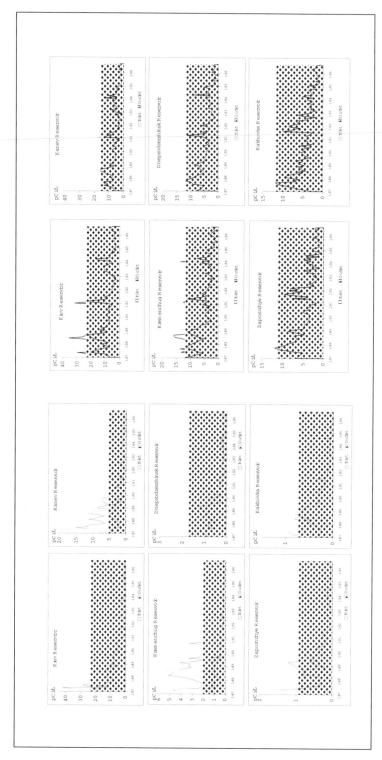

**FIGURE 5.3.4.** Variation in mean monthly concentrations of (a) $^{137}$Cs and (b) $^{90}$Sr in the water at the inlet and outlet of the reservoirs 1987– 1996, $10^{-9}$ Ci/L

Research performed since the accident suggests that approximately 15% of the $^{137}$Cs and 45% of the $^{90}$Sr contamination transported by the Pripyat and Dnepr Rivers into the Kiev Reservoir came from the Exclusion Zone. This contribution is not constant, but varies from season to season, depending on the water flow level for the particular year in question, how the high water moves through the Exclusion Zone, the flooding conditions in the contaminated floodplain, the snow level in the Exclusion Zone, and the snowmelt rate. Calculations based on the UkrNIGMI balance surveys indicate that at least 70% of the $^{137}$Cs influx into the Pripyat River is from a broad area of the watershed in the Ukrainian and Belarussian Polessye. This value is similar to that independently obtained from the observational data by Ukrainian Center for Environmental Monitoring (77%). This is also indirectly confirmed by similar specific activities of suspended solids in the Upper Dnepr (whose watershed does not include any sections of the Exclusion Zone) and Pripyat Rivers. The mean specific activity of Pripyat River suspended solids transported into the Kiev Reservoir between 1987 and 1992 is 4 kBq/g, while that of Upper Dnepr suspended solids was 3.5 kBq/g. From June 1986 to December 1995, approximately 3,800 Ci of $^{137}$Cs and 4,200 Ci of $^{90}$Sr were transported into the upper reaches of the Dnepr reservoir system by various rivers. The Pripyat River accounted for 66% of the total influx of $^{90}$Sr; note that the Pripyat River contribution increased from 49% to 75% from 1987 to 1994.

The annual external $^{137}$Cs and $^{90}$Sr balance (inflow vs. outflow) for the Dnepr reservoir system was calculated using a large volume of radiation monitoring data that was prefiltered with respect to certain data quality criteria. The inflow side of the balance includes the waterborne radionuclides flowing into the reservoir, while the outflow side includes the waterborne radionuclides discharged through the hydroelectric power plant. The difference between the inflow and outflow is equal to the radionuclides retained in the reservoir. This value includes the accumulation of radionuclides in bottom sediments, the number of radionuclides in the water at the end of the period, and the number of radionuclides removed with water for household, industrial, and agricultural use. Data on the external radionuclide balance can be used to characterize the accumulation capacity of the reservoirs given the radioactive contamination that occurred in 1986 and early 1987 (Figure 5.3.5).

The waterborne $^{137}$Cs concentration decreases by more than an order of magnitude from Chornobyl to Svetlovodsk. The waterborne concentration gradually decreases from Svetlovodsk to Zaporozhye. The $^{137}$Cs concentrations in the Kakhovka Reservoir have been at the pre-accident level since 1993. Sedimentation is particularly rapid in the Kiev and Kanev Reservoirs, which annually trap approximately 75% of the $^{137}$Cs coming into the reser-

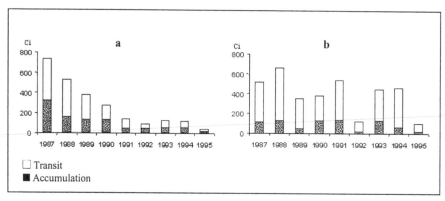

**FIGURE 5.3.5.** Radionuclide balance in Kiev Reservoir: (a) $^{137}$Cs; (b) $^{90}$Sr, Ci

voir system. The highest $^{137}$Cs accumulation rate is observed in the Kremenchug Reservoir—77%. The Kakhovka Reservoir also provides good conditions for $^{137}$Cs sedimentation, although observations indicate that virtually no cesium radionuclides reach the lower Dnepr Reservoir, having been intercepted by the upper reservoirs.

A factor of 6–10 decrease in $^{90}$Sr concentrations from the Pripyat River near Chornobyl to the Kakhovka Reservoir at Kakhovka is largely due to dilution of the Pripyat River water by uncontaminated water from various tributaries. During June 1994 measurements of the waterborne radioactive contamination in the reservoir system, elevated concentrations of $^{90}$Sr were observed in the Dneprodzerzhinsk Reservoir, which at that time contained a contaminated mass of water that had formed in the Upper Dnepr River in February–April (Figure 5.3.6).

By September 1994, this contaminated mass of water that had formed in the Upper Dnepr River filled the upper portion of the Kakhovka Reservoir. The ameliorative capacity of the Dnepr reservoir system for waterborne $^{90}$Sr involves equalization (reduction) of maximum (peak) concentrations and delaying the passage of contaminated water masses. Even very high peak $^{90}$Sr concentrations in the Pripyat River are smoothed out by the time the Kremenchug River is reached, and by the time the Kakhovka Reservoir is reached, only gradual increases and decreases in the concentration of this radionuclide are observed, reflecting the water-mass transformation process in the reservoir system.

Waterborne radioactivity monitoring data from the time of the Chornobyl NPP accident until December 1995 indicate that at least 222 x $10^{12}$ Bq of $^{137}$Cs and 185 x $10^{12}$ Bq of $^{90}$Sr entered the Dnepr reservoir system from atmospheric fallout and runoff. Over this same time period, the Dnepr system discharged approximately 1.48 x $10^{12}$ Bq and 52 x $10^{12}$ Bq of accident-related $^{137}$Cs and $^{90}$Sr, respectively, into the Black Sea. Thus, the

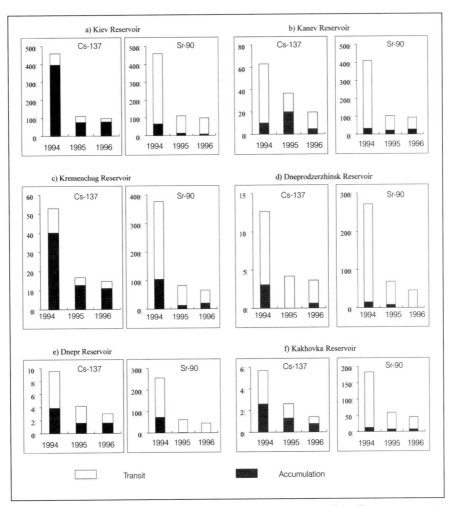

**FIGURE 5.3.6.** Radionuclide balance in various reservoirs of the Dnepr reservoir system, 1994–1996, Ci

ecosystem formed by the Dnepr reservoirs and the Dnepr-Bug lagoon retained at least 99% of the $^{137}$Cs and 70% of the $^{90}$Sr that entered these bodies of water from the accident. Most of the retained radionuclides were deposited in sediments at the bottom of the reservoirs. Since 1993, all of the $^{137}$Cs from rivers originating in contaminated watersheds have been trapped in the Dnepr reservoir system.

The sedimentary water treatment process in the Dnepr reservoirs is largely based on suspended particles that adsorb radioactive material and other pollutants onto their surfaces; the particles settle out along the length of the reservoir, taking the pollutants with them. For example, the Kiev Reservoir retains more than 90% of the solid matter entering the reservoir. The main source of this material is suspended matter in river discharges (74%). Assessments of the sedimentary balance in the Kiev Reservoir [Tertyshnik et al. 1980] suggest that no more than 6% of the suspended matter is capable of transiting the reservoir (on an annual average basis), making up 74% of the solid matter discharged from the Kiev and Kanev Reservoirs (66,000 metric tons). The remaining 26% consists of phytoplankton and shore erosion products. Typical data describing the adsorption of $^{137}$Cs on suspended particles of various sizes are provided in Figure 5.3.7. The scatter is due to seasonal variations in particle composition and in the physical and chemical characteristics of the ambient water. Approximately 50% of the suspended $^{137}$Cs is deposited in the Kiev Reservoir following adsorption onto clay particles. Only particles smaller than 5 $\mu$m in diameter are capable of transiting the Kiev Reservoir at the typical flow velocity of 2–5 cm/s, while most (up to 73%) particles with hydraulic sizes larger than this will be deposited on the bottom.

These data indicate that the Kiev Reservoir is the most important accumulation site for suspended particles and associated radionuclides in the Dnepr reservoir system. This geochemical process also affects the tendency for the radioactive contamination in the reservoir to vary with suspended matter transport distance and time required for $^{137}$Cs to accumulate in the bottom deposits of the reservoir (Figure 5.3.8).

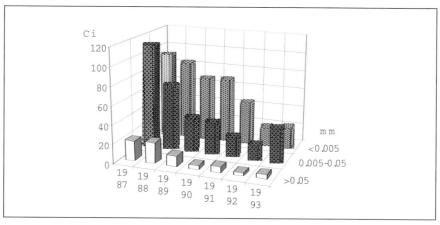

**FIGURE 5.3.7.** Removal of $^{137}$Cs by suspended matter fractions in Kiev Reservoir (1987–1993)

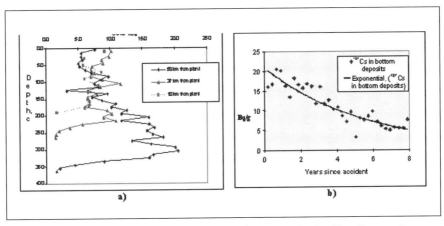

**FIGURE 5.3.8.** Vertical distribution of specific activity in the Kiev Reservoir
(a) various distances from the Kiev Hydroelectric Power Plant
and (b) exponential trend for formation of contaminated layers
in bottom sediments as a function of time since the accident

Little work has been done on certain suspended matter formation
processes in rivers and reservoirs because they are episodic in nature, and
there are no well-defined methodological approaches for studying them.
Processes in this category include the formation of weakly soluble iron and
manganese hydroxides, which have a high formation rate during the ice-cov-
ered period, when the oxygen content of the water is low. Such compounds
have been shown to be capable of binding dissolved [137]Cs. The presence of
high-molecular-weight compounds such as humic acid and fulvic acid may
be an important factor in the redistribution of [137]Cs between the liquid and
solid phases.

### 5.3.6 Radioactive Contamination of Bottom Sediments in Dnepr Reservoir System

Radioactive [137]Cs contamination of the bottom sediments essentially
ended in 1988, when the amount radioactive runoff in the major tributaries
noticeably decreased. Between 1990 and 1994, UkrNIGMI performed a survey of the bottom sediments in the Dnepr reservoir system. This survey has
been used, in conjunction with analyses of the bottom samples, to construct
contour maps of the [137]Cs distribution on the bottom of the Kiev, Dneprodzerzhinsk, and Kakhovka Reservoirs, and maps for the Kremenchug
and Kakhovka Reservoirs are currently being constructed [Kashparov et al.
1997b]. These same samples were also used to determine [90]Sr concentrations. However, a problem was encountered when mapping the spatial

distribution of $^{90}$Sr over the bottom of the reservoirs. Although a bathymetric chart was used to map the spatial distribution of $^{137}$Cs, no similar basis is currently available for constructing maps of the $^{90}$Sr distribution.

A map of the spatial distribution of mollusks and shells could be used as the basis for a $^{90}$Sr map, as the highest concentrations of $^{90}$Sr are observed in mollusk valves. Permanent retention of $^{90}$Sr in [mollusk] valves may be one of the primary factors in the irreversible removal of this radionuclide from the water into the bottom.

As riverborne radionuclide transport continues to decrease, it becomes easier to follow the process of radionuclide redistribution over the bottom. This process involves various hydrodynamic factors (inflow currents, drift currents, and the effects of storms on the bottom) that cause finely dispersed particles with high specific contamination levels to be removed from shallow areas and redeposited in stable sedimentation areas, such as the old Dnepr River bed, depressions on the submerged floodplain (lakes in the floodplain, open oxbows, and closed oxbows), and areas along the shore of the reservoir where finely dispersed suspended particles eroded from the shore accumulate (at the bases of underwater shore slopes). These are precisely the areas where the bottom deposits are observed to have the highest density of radionuclide contamination in all of the reservoirs. Analysis of the vertical distribution of $^{137}$Cs in the bottom deposits of the various reservoirs indicates that in areas of stable sedimentation, the layer with the highest contamination (dated 1986) is gradually being buried. For example, in deep areas of the Kakhovka and Kremenchug Reservoirs, the layers with the highest contamination levels, dated 1986–1987, are 20 cm or more below the surface of the bottom, indicating that the deposition rate in these areas is high. The most contaminated, silty deposits are localized in areas of deep water and naturally become buried under a layer of less-contaminated sediment. These two factors further reduce the likelihood of the radionuclide transport into the water and have a favorable effect on water quality throughout the reservoir system. Because $^{137}$Cs can be permanently sorbed onto finely dispersed mineral and organic matter, it does not diffuse or infiltrate into the bottom.

Table 5.3.6 lists the amount of $^{137}$Cs contained in bottom deposits throughout the Dnepr reservoir system, along with the mean contamination densities of the bottom as of 1 January 1995.

Another important question is whether or not the bottom deposits in the reservoir system are currently a secondary source of $^{137}$Cs contamination. Radiation monitoring of the water answers this question in the negative. Calculations of the external radionuclide balance indicate that each of the reservoirs, as well as the reservoir system as a whole, serves as a sink for deposition of radionuclides into the bottom deposits. The reverse process—

**TABLE 5.3.6.** $^{137}Cs$ in Dnepr Reservoir System Bottom Deposits (1995)

| Reservoir | Total Content, $10^{10}$ Bq | Contamination Density, $10^4$ Bq/m$^2$ |
|---|---|---|
| Kiev | 7955±555 | 8.61±0.59 |
| Kanev | 3700±1480 | 5.48±2.55 |
| Kremenchug | 2627±259 | 1.18±0.11 |
| Dneprodzerzhinsk | 426±44 | 0.74±0.07 |
| Dneprovsk | 222±74 | 0.55±0.18 |
| Kakhovka | 610±74 | 0.30±0.04 |
| System Total | 15540±2480 | |

remobilization of $^{137}Cs$ from the bottom deposits into a layer of water near the bottom via a pore solution with a higher radionuclide concentration than the main body of water—undoubtedly exists as well. However, the current characteristic $^{137}Cs$ remobilization rate is an order of magnitude smaller than the characteristic fixation rate, corresponding to a resultant net incorporation of $^{137}Cs$ into the bottom via sedimentation, mollusk bioaccumulation, etc. The resultant $^{90}Sr$ flow is also directed toward incorporation in bottom deposits, despite the much higher mobility of strontium in the reservoirs.

Secondary contamination of the water through liberation of soluble radionuclide forms due to disruption of bottom deposits also seems unlikely at this point. The self-treatment process in shallow areas periodically subject to wave action has run to completion, and the bulk of the $^{137}Cs$ is now concentrated in areas of deep water. Measurements performed on water samples and samples of suspended solids collected (by the authors) during the most intense storm on the Kanev Reservoir within the past 10 years (25 May 1992) showed that the sharp increase in suspended matter because of the storm did not lead to any appreciable increase in the radionuclide concentration, either in the suspended matter or in solution.

In our view, the increased seepage from reservoirs when extreme high-water events pass through them will not lead to radionuclide washout in areas of maximum contamination. The unprecedented drawdown of the Kanev Reservoir on 26–30 April 1986 (when 80% of the total volume of the reservoir was discharged into the tailrace) did not noticeably disturb the bottom deposits, and confirmed that the silty deposits were resistant to hydraulic action. In the most unfavorable case, the open and closed oxbows where the Pripyat and Dnepr merge might be scoured. However, the inevitable slowing of the flow when it reaches the main body of the Kiev Reservoir will cause most of the scoured material to settle to the bottom, and releasing the water-soluble forms of the radionuclides will not lead to a critical increase in the waterborne radionuclide concentration.

Thus, under current conditions, riverborne transport of radionuclides from accident-contaminated watersheds at the head of the Dnepr reservoir system is the only substantial source of radioactive contamination for the entire Dnepr reservoir system. The total $^{137}$Cs and $^{90}$Sr runoff and, thus, the concentration of these radionuclides in the incoming water, is clearly decreasing as a function of time.

As water masses move down from the upper reaches of the Dnepr water system to the Black Sea, the radionuclide concentrations decrease as a result of dilution with water from uncontaminated tributaries and sedimentation. As a result, there is a decrease of over two orders of magnitude in $^{137}$Cs concentration from Chornobyl to Zaporozhye, where the $^{137}$Cs concentration becomes equal to the background value. The decrease in $^{90}$Sr concentration is equal to approximately one order of magnitude. The concentrations of this radionuclide observed at the main irrigation intakes over the past 3 years have not exceeded the established threshold value for irrigation water— 259 Bq/L.

Contaminated water masses from the upper Dnepr watershed currently lose nearly 100% of their $^{137}$Cs and approximately 70% of their $^{90}$Sr as they travel through the entire Dnepr reservoir system to the Black Sea; these radionuclides accumulate in bottom sediments.

The suspended matter from the Pripyat and Upper Dnepr generally consists of a mineral component that was contaminated in its initial location, and therefore is a negative factor affecting water quality in the upper Dnepr watershed. As the water moves through the reservoirs, this suspended matter settles to the bottom; the settling time is determined by the hydraulic size. The nature of the physical interaction between the $^{137}$Cs and the mineral component prevents any appreciable desorption of the radionuclide into solution over a wide range of physical and chemical indicators used to describe the ambient water. Thus, the water mass is "naturally treated" through deposition of suspended solids (which were initially contaminated at their place of origin) on the bottom of the reservoirs.

Autochtonous suspended matter—uncontaminated material from shore erosion and phytoplankton—has a high sorption capacity. They actively absorb dissolved radionuclides from the water and treat the water solution. Subsequent sedimentation of this suspended matter completes the cycle for most radioactive materials in the water.

Mechanical differentiation of the solid material on the bottom by hydraulic size causes the most highly contaminated suspended matter to be concentrated in areas of deep water unaffected by hydraulic effects. The reduction in riverborne radionuclide transport has given rise to a steady decrease in the radionuclide content of bottom deposits throughout the reservoir system due to radioactive decay. The likelihood of appreciable secondary contamination of the Dnepr is extremely low.

Extremely high water in the Pripyat and Dnepr watershed may cause temporary increases in $^{137}$Cs and $^{90}$Sr transport rate. However, such increases will consist of a relatively brief, isolated pulse and will not impact the general trends affecting the radiation environment in the Dnepr watershed.

The bulk of the primary source data reviewed in this section was obtained by the following organizations: UkrNIGMI, Goskomzhilkommunkhoz, Goskomvodkhoz, Ukrainian Ministry of Health, UDK Pripyat, and Taifun.

## 5.4 RADIOACTIVE CONTAMINATION OF GROUNDWATER

### 5.4.1 Groundwater in the Exclusion Zone

In analyzing the data for groundwater within the Exclusion Zone, we feel the data on subsurface water contamination in the zone from 1986–1992 are incomplete and unreliable, because of imperfections in the monitoring network and methodological errors in the sample collection process. Expansion and improvement of the subsurface water monitoring system is currently the most important hydrogeological research issue for the Exclusion Zone.

The hydrogeological migration of radionuclides from interim radioactive waste storage and disposal sites and the Chornobyl NPP cooling pond was first detected in late 1988 and early 1989, and became widespread in 1990–1992. Since that time, radioactive contamination of subsurface water has been considered a serious radioenvironmental problem in the Exclusion Zone.

By 1995–1996, migration from the surface had led to substantial $^{90}$Sr and $^{137}$Cs contamination (in concentrations 2–3 orders of magnitude higher than the pre-accident background level) in the upper section of the groundwater aquifer in the inner Exclusion Zone (5–10 km). In areas where radioactive waste is buried near the surface (storage and disposal sites, Chornobyl NPP site, etc.), the $^{90}$Sr groundwater concentration varies from $10^3$–$10^4$ Bq/L.

The artesian aquifer in the Eocene deposits, which serves as the source of potable water within the Exclusion Zone, is only beginning to become contaminated. However, in view of the importance of this aquifer as a water supply, improvement of the radioenvironmental monitoring system for this aquifer should be a high-priority task.

Hydrogeological migration of $^{90}$Sr, which has high geochemical mobility, is an important issue. While the radiological environment has stabilized or improved in other respects, predictions indicate that radioactive contamination of groundwater will increase over the next few decades.

5.4.1.1 **Hydrological and Geological Conditions in Exclusion Zone.**
The aquifers in the Exclusion Zone lie in a rapidly replenished hydrodynamic zone for which atmospheric precipitation is the main water source. The local recharge areas are along high-elevation watersheds. The groundwater is mainly discharged into the Pripyat River and a network of streams and drainage channels (maximum density of dissected terrain and drainage is 0.3 km/km$^2$).

The groundwater (i.e., the first aquifer from the surface) is uniformly distributed and confined to Quaternary sediments. In excess of approximately 30% of the Exclusion Zone (by area) has groundwater at a depth of 0–1 m, more than 50% of the area has groundwater at 1–3 m, 10% of the area has groundwater at 3–5 m, and 10% of the area has groundwater at more than 5 m. The aquifers are from 15–30 m in thickness. The water production capacity of the aquifers is 100–400 m$^2$/d in fluvial terraces and 40–100 m$^2$/d in watershed surfaces. The seasonal water-level amplitude is 0.5–1 m along watersheds and 1–2 m or more in floodplains. The recharge infiltration level is 50–200 mm/yr at typical precipitation levels of approximately 600 mm/yr. Average regional groundwater head gradients run from 1 x 10$^{-3}$–5 x 10$^{-3}$. Actual groundwater velocities (assuming an effective porosity of order 0.2 for the sandy deposits) over most of the Exclusion Zone are estimated to be 10–50 m/yr. Calcium and magnesium bicarbonate are the major impurities, with a total mineral content of 0.1–0.5 mg/L.

The Kiev Eocene marl formation (P$_2$kv) is the first regional-scale weakly permeable layer separating the groundwater from the glauconite/quartz-sand artesian aquifer in the Buchak and Kanev Eocene formations (P$_2$bc). The marl ranges from 10–20 m in thickness, with occasional "holes" that are filled with sandy deposits. The estimated filtration coefficients range from 2.5 x 10$^{-4}$–2 x 10$^{-2}$ m/d. The artesian aquifer in the Buchak and Kanev Eocene formations lies at a depth of 20 m in the western portion of the Exclusion Zone, and 90 m in the northeastern portion of the Exclusion Zone and under morainic ridges. Chornobyl NPP uses this layer for its water supply. This layer is underlain by several artesian aquifers in chalk deposits that have been studied down to the underlying Jurassic clay at a minimum depth of approximately 220 m.

Thus, hydrogeological conditions predispose the Exclusion Zone to radionuclide migration from the soil surface into the groundwater. The high humidity, low slopes (with the associated low surface runoff levels),  soil characteristics (weakly humic acid-reacting derno-podzolic soils), and lithological composition of the topsoil (primarily quartz sand, which has low sorption properties) all facilitate migration of radionuclides released in the Chornobyl NPP accident. The first, nonartesian aquifer in Quaternary sediments (groundwater), which had been widely used by the rural population for drinking water before the accident, was the most severely affected by radioactive contamination.

### 5.4.1.2 Subsurface Water Monitoring System.

Very little data are available on the pre-accident background radioactive contamination of natural water in the vicinity of Chornobyl NPP. Isolated references in the literature suggest that the concentrations of $^{90}$Sr and $^{137}$Cs in groundwater at the Chornobyl NPP site before the accident were on the order of $10^{-3}$–$10^{-2}$ Bq/L. In many cases, the lack of any reliable data from pre-accident background monitoring makes it more difficult to interpret post-accident observations.

Data on subsurface water contamination in the vicinity of Chornobyl NPP in 1986–1987 are also sketchy. This is because the observational data obtained by various organizations (mainly at the all-Union level) in 1986–1987 were not systematically organized and collected at a single center. As a result, much of the data from the first years following the accident have apparently become "buried" in the archives of various Russian institutes or have simply been lost.

GP RADEK (successor to the Dosimetric Monitoring Administration) is currently conducting groundwater observations at approximately 100 boreholes (15–30 m deep) and Eocene artesian aquifer observations at approximately 20 boreholes (80–100 m deep). These numbers include on-site facilities at radioactive waste disposal sites. Samples are collected from each of the wells on a monthly to quarterly basis. Most of the wells are concentrated toward the center of the inner Exclusion Zone (5–10 km).

The data from the "old borehole network" are unreliable, unfortunately. Ever since 1986, the reliability of the observations has been adversely affected by three factors: contamination of the boreholes with radionuclides from the surface soil during the drilling process, failure to pump the boreholes before sample collection, and poor design and materials used in the borehole casings.

Each hydrogeological monitoring station should ideally consist of a multilevel battery of boreholes made from a hydrochemically inert material (plastic) with a 1- to 2-m-long filter that samples a discrete segment of the aquifer.

The hydrogeological monitoring effort has not been sufficiently comprehensive. Systematic monitoring of water levels under the Exclusion Zone was only begun in 1990. There is no routine geochemical monitoring of the subsurface water. Standard recommendations developed on the international level also call for recording easily determined in situ parameters such as pH, electrical conductivity, and water temperature, which significantly enhance the data content of the observations.

5.4.1.3  **Time Dependence of Geological Migration Processes.** Hydrogeological migration of radionuclides is generally a relatively slow process governed by the sorption characteristics of the subsoil, as well as aeration-zone imperviousness and perching. Gradual leaching of radionuclides from the Chornobyl fuel matrices is another factor that affects radionuclide migration. The time required for half the radionuclides to be leached from the fuel particles is estimated to range from 1–2 years to 7–10 years depending on size and physico-chemical characteristics.

The long-lived radionuclides in the Chornobyl release can be ranked as follows with respect to geological mobility:

$$Sr > Cs > Am > Pu.$$

Sorption is generally characterized in terms of the distribution coefficient $(K_d)$, which is the ratio of the equilibrium concentrations of the radionuclide in the solid phase and in solution (with higher $K_d$ corresponding to higher sorption). Research performed by the Radium Institute (St. Petersburg), the Institute of Geological Sciences, and other organizations indicates that the sorption distribution coefficients between subsurface water and Quaternary sand deposits for $^{90}Sr$ are generally 1–10 mL/g [Gudzenko et al. 1994]. Sorption is an ion-exchange process. For conservative $^{90}Sr$ geological migration predictions (i.e., predictions that exaggerate the amount of migration), a distribution coefficient of 0.5 mL/g can be recommended; for "standard" predictions, a distribution coefficient of 2–4 mL/g is recommended. For $^{137}Cs$, the distribution coefficient generally ranges from 10–100 mL/g. The low mobility of $^{137}Cs$ favors absorption of the radionuclide by the subsoil matrix by mechanisms other than ion exchange ("fixation" by clay minerals). Less research has been performed on the sorption characteristics of long-lived transuranic radionuclides, which typically show complex geochemical behavior. Radium Institute laboratory tests on sedimentary rocks from the Ryzhii Les Interim Radioactive Waste Storage Site led to values of 100–1,000 mL/g for the distribution coefficients of the plutonium isotopes and $^{241}Am$. These results indicate that plutonium and americium have low migration rates. Thus, hydrogeological migration of $^{90}Sr$, which has a low sorption capacity in rock, poses the greatest risk. In addition, the maximum permissible concentration of $^{90}Sr$ in drinking water is extremely low.

The extremely slow nature of the geological migration process meant that the issue of radioactive contamination in subsurface water from the Exclusion Zone was not immediately apparent, and attitudes regarding this issue have evolved significantly. During the months immediately following the accident, the risk of subsurface water contamination was overstated. During the summer of 1986, several protective drainage systems were installed (a total of approximately 300 boreholes were drilled) to protect the subsurface water and prevent geological migration of radionuclides into the Pripyat River and low-lying aquifers in the inner Exclusion Zone (5–10 km). During the same summer, construction of an underground clay "wall" around Chornobyl NPP was begun. By late 1986, when more accurate data on the surface contamination had become available, a low radionuclide leach rate for the fuel particles was proven in laboratory tests, and monitoring had not shown any radionuclides in subsurface water, these protective measures were recognized as excessive. Construction was stopped on the underground "wall" and the drainage systems were placed in "wait status" and then partially dismantled.

According to data obtained by the Dosimetric Monitoring Administration, the first traces of radionuclides in groundwater from the Exclusion Zone were detected after the spring snowmelt in April–May 1987. The main radionuclides detected were $^{134,137}$Cs, $^{106}$Ru, and $^{90}$Sr in concentrations 2–3 orders of magnitude smaller than the maximum permissible concentrations. Through 1988, the concentrations of the short-lived radionuclides ($^{134}$Cs and $^{106}$Ru) decreased substantially, while the $^{90}$Sr concentrations increased, especially for the boreholes located near the Chornobyl NPP cooling pond.

The problem of radioactive contamination in subsurface water became much more significant in late 1988 and early 1989. In October 1988, the Dosimetric Monitoring Administration found that $^{90}$Sr in the groundwater from PK-113, a monitoring location in the cooling pond levee, was consistently exceeding the maximum permissible concentrations. In the fall of 1989, All-Union Scientific Research Institute for Geological and Non-Ore Minerals performed the first hydrogeological tests at the Ryzhii Les Interim Radioactive Waste Storage Site, and detected $^{90}$Sr and $^{239,240}$Pu in the groundwater from flooded radioactive waste trenches at concentrations of $10^2$–$10^3$ Bq/L and on the order of 1 Bq/L [Afonin et al. 1992]. By 1990–1992, the process of hydrogeological migration had taken hold at the interim radioactive waste storage site and cooling pond, and the issue of radioactive contamination in subsurface water was beginning to be considered a serious radioenvironmental problem in the Exclusion Zone.

The limited nature of the routine monitoring performed using the official stationary borehole network required special groundwater monitoring efforts using manually drilled temporary boreholes. In 1993–1995, manual drilling performed by Geological Sciences Institute indicated that virtually the entire first aquifer from the surface throughout the inner Exclusion Zone (5–10 km) was contaminated with $^{90}$Sr in excess of interim maximum permissible levels (1991), indicating that $^{90}$Sr had migrated from the surface through the aeration zone. Special hydrogeological studies—performed by Geological Sciences Institute and Science and Engineering Center Koro—

at the Ryzhii Les, Stroibaza, Neftebaza, and Peschanoe Plato Interim Radioactive Waste Storage Sites, as well as routine monitoring performed by Dosimetric Monitoring Administration using the stationary borehole network indicated that the groundwater was gradually being contaminated with $^{90}$Sr. (For more information on the waste storage sites, see Chapter 4).

### 5.4.1.4   Radioactive Contamination of Subsurface Water in 1995–1996.
The sources of radioactive subsurface water contamination in the Exclusion Zone can be arbitrarily classified into two categories: "point sources" and "distributed sources." The most important point sources are as follows: the Shelter and adjacent areas of the Chornobyl NPP site, the cooling pond, and the interim radioactive waste storage sites. The distributed sources include the surface of soil contaminated with radioactive fallout, where migration occurs with the infiltration of water through the aeration zone. While the point sources create localized "halos" of radioactive subsurface contamination, the distributed sources contaminate the subsurface water on a regional level. Table 5.4.1 summarizes the available data on radionuclide contamination of groundwater as of 1995–1996. Most of the available data on radioactive contamination of subsurface water is for the inner Exclusion Zone (5–10 km).

Because of the limited nature of the State Enterprise for Radiation, Dosimetric and Environmental Monitoring monitoring network and methodological deficiencies in the data obtained using the official Geological Sciences Institute borehole network in 1993–1996, special monitoring of subsurface water was performed using manually drilled boreholes. During 1995–1996,

**TABLE 5.4.1.**   Typical Radioactive Contamination Levels for Subsurface Water in the Inner Exclusion Zone (5–10 km) as of 1995–1996

| Source | $^{90}$Sr (MPC = 14.8 Bq/L) | $^{137}$Cs (MPC = 555.5 Bq/L) | $\Sigma^{239,240}$Pu (MPC = 81.5 Bq/L) | $^{241}$Am (MPC = 70.3 Bq/L) |
|---|---|---|---|---|
| Regional groundwater contamination | 0.1–10 | 0.1–1 ≤0.01 | ≤0.01 | |
| Shelter | 10–10$^3$ | 0–100 | — | — |
| Interim radioactive waste storage sites | 10$^2$–10$^4$ | 0.1–10 | ≤0.1–1 | ≤0.1 |
| Cooling pond | 1–10$^2$ | 0.1–1 | — | — |
| Artesian aquifer in Eocene deposits | ≤0.01 | ≤0.01 | — | — |

Data provided by Geological Sciences Institute, State Enterprise for Radiation, Dosimetric and Environmental Monitoring, the Institute for Nuclear Research, and Science and Engineering Center Koro.

MPC = maximum permissible concentration.

Geological Sciences Institute established a regional network of manually drilled, small-diameter (1-in.) plastic boreholes throughout the inner Exclusion Zone (5–10 km). These boreholes were employed to sample the top 1 m of the aquifer. The $^{137}$Cs groundwater concentration was found to be generally stable, with typical values on the order of a few becquerel per liter or a few tenths of a becquerel per liter, and are generally well below the maximum permissible concentrations. The observational data demonstrate that migration from the soil surface has led to $^{90}$Sr and $^{137}$Cs contamination of the upper section of the aquifer in the inner Exclusion Zone (5–10 km) to a level that is two to three orders of magnitude higher than the pre-accident background level. In areas where radioactive waste is buried near the surface (interim radioactive waste storage sites, Chornobyl NPP site, etc.), the $^{90}$Sr groundwater concentration varies from $10^3$–$10^4$ Bq/L, i.e., there are distinct areas within which the groundwater is contaminated with $^{90}$Sr. Predictions suggest that the groundwater contamination will continue over the next several decades as $^{90}$Sr continues to leach out of the fuel fallout and migrate through the aeration zone into the aquifer.

The low specific activities of plutonium and americium in Exclusion Zone groundwater are not currently cause for alarm; the only exception involves the migration of transuranic elements in the geological filtration medium of the Shelter, where specific conditions (both technical and radiochemical) and the extremely high activities mean that migration of transuranic elements may lead to additional risk.

The artesian aquifer in the Eocene deposits is only beginning to become contaminated. The $^{90}$Sr and $^{137}$Cs concentrations in the subsurface water extracted by the Pripyat water intake are generally on the order of a few hundredths of a becquerel per liter.

### 5.4.2 Groundwater in Areas Adjacent to Exclusion Zone

Radionuclide contamination is beginning to appear in the subsurface water used for water supplies in areas adjacent to the Exclusion Zone. The most complete studies of this subsurface water have been performed in the Kiev Metropolitan Area. This area includes the city of Kiev, its suburbs (Vyshgorod, Irpen, Boyarka, Brovary, Borispol, etc.), and land within a 60–70 km radius. The aquifers and aquifer complexes in this area are associated with Quaternary, Eocene, Cenomanian-Callovian, and Middle Jurassic deposits (Figure 5.4.1).

Studies of the groundwater in areas adjacent to the Exclusion Zone show that downward vertical radionuclide migration paths are a fundamental element in aquifer contamination. Lateral migration plays a secondary role in regional subsurface water contamination processes. In addition to the

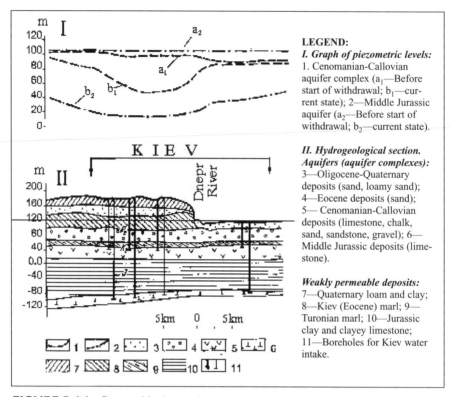

**FIGURE 5.4.1.** Regional hydrogeological section for the city of Kiev

natural radionuclide migration pathways, we have experimentally confirmed the existence of artificial pathways due to imperfections in and around the boreholes. (For more information, see Section 5.4.2.2.)

Having established migration pathways, a variety of other factors determine the subsurface water contamination rate. These factors include the density and type of fallout, landscape and geochemical conditions, sorptive capacity and homogeneity of geological medium, and intensity of filtration connections between various elements of the geological water circulation system. Correlations were established between the subsurface water recharge rate, the depth of the subsurface water depression, the thickness of the weakly permeable Kiev marl layer, the sorption characteristics of the soils, the surface density of radionuclides and the $^{137}Cs$ and $^{90}Sr$ concentration under natural conditions and under conditions of rapid water drawdown. (For more information, see Section 5.4.2.3.)

Despite the initial contamination of subsurface water with Chornobyl radionuclides, artesian aquifers still remain the most reliable source of water within the affected regions. A comprehensive monitoring system will be required for reliable prediction and, if necessary, management of water quality. This monitoring system will need to include a broad, reliable system of wells and analyses. (For more information, see Section 5.4.2.4.)

### 5.4.2.1 Hydrological and Geological Conditions Adjacent to the Exclusion Zone.

In the area adjacent to the Exclusion Zone, the aquifer complex in the Cenomanian-Callovian deposits and the aquifer in the Middle Jurassic (Bajocian) formations are the most important for large-scale centralized water supply systems and are widely used. The quality of these aquifers is largely determined by that of the recharge aquifers (primarily the aquifers nearest the surface).

Analysis of several years' observations of hydrodynamic and hydrochemical conditions within the Kiev Metropolitan Area indicates that the volume and depth of water exchange are increasing as a result of human activities.

An artificial hydrochemical regime develops around the center of the water depression in the aquifer being used. This regime is characterized by the following:

- High spatial variability in the macro-chemical composition of the subsurface water being used

- Increased temporal variability in the chemical composition of the subsurface water being used (relative to similar aquifers not being used).

However, we can generally state that these spatial and temporal variations are limited in nature, and have not substantially affected the quality of the subsurface water. Note, the oxidizability of the subsurface water is constant in space and time, indicating that there are no significant organic pollutants in the subsurface water.

The accident at Chornobyl NPP led to contamination of the subsurface water in the first aquifer from the surface, as well as other, deeper aquifers, both within the Exclusion Zone and at substantial distances from the plant. Almost all of the water samples collected showed measurable concentrations of $^{137}Cs$ and $^{90}Sr$.

The following subsurface water test depths were used: 2–18 m (Quaternary aquifer), 45–65 m (Eocene aquifer), 80–150 m (Cenomanian-Callovian aquifer), and 200–300 m (Bajocian aquifer).

A total of 600 $^{137}Cs$ and 400 $^{90}Sr$ determinations were performed from 1992–1996. The measured concentrations of both radionuclides ranged from a few millibequerels per liter  to hundreds of millibequerels per liter (Table 5.4.2).

**TABLE 5.4.2.** Percentages of Measured Concentrations for Radionuclides in Kiev Metropolitan Area Subsurface Water

| Age of Aquifer | [137]Cs Concentration, mBq/L | | | | [90]Sr Concentration, mBq/L | | | |
|---|---|---|---|---|---|---|---|---|
| | <10 | 10–50 | 51–150 | >150 | <10 | 10–50 | >50 |
| Quaternary | 24% | 56% | 14% | 6% | 53% | 40% | 7% |
| Eocene | 45% | 41% | 9% | 5% | 71% | 21% | 8% |
| Cenomanian-Callovian | 46% | 31% | 13% | 10% | 80% | 20% | — |
| Bajocian | 47% | 28% | 15% | 10% | 62% | 35% | 3% |

As an example of the spatial distribution of radionuclide contamination in subsurface water, Figures 5.4.2 and 5.4.3 show the distribution of [137]Cs in the Kiev-region Quaternary and Eocene aquifers.

Several specific instances where [134]Cs was found in water samples indicate that the radionuclides originated from Chornobyl. In this connection, it should also be emphasized that control samples of subsurface water obtained under similar hydrogeological conditions but outside of the contaminated area (less than 20 kBq/m²) did not show any significant quantities of these radionuclides.

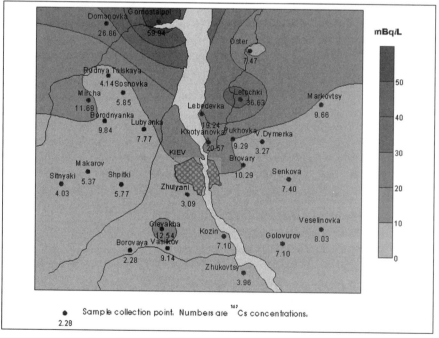

**FIGURE 5.4.2.** Distribution of [137]Cs in the Quaternary aquifer under the Kiev Metropolitan Area

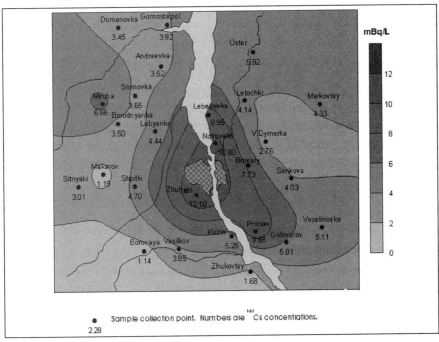

**FIGURE 5.4.3.** Distribution of $^{137}$Cs in the Eocene aquifer under the Kiev Metropolitan Area

5.4.2.2 **Migration Pathways.** It is very important to determine the pathways by which these radionuclides entered the aquifers. Downward migration of radionuclides has been found to play an important role in the contamination of multilayer aquifers. Lateral migration plays a secondary role in regional subsurface water contamination processes because of the low rate of lateral filtration.

In addition to lateral radionuclide migration, artificially induced radionuclide-penetration mechanisms (associated with imperfections in and adjacent to boreholes) have also been identified. The contribution of artificial migration to the overall radioactive contamination is small, but it could lead to significant distortion of radionuclide concentrations measured during episodic groundwater testing.

Of course, we should also note that there are wells near Kiev which have accessed Eocene deposits at depths of 40–60 m with measurable concentrations of $^{134}$Cs and $^{137}$Cs and are known not to have artificial migration routes because they are artesian wells. The location of these wells suggests that the downward radionuclide migration rate must be in excess of 10–15 m/yr. By extrapolating the possible natural downward migration of these radionuclides to the Jurassic level at depths of up to 250 m, we obtain migration rates of 50 m/yr or more in the vicinity of a depression.

These high migration rates suggest that the so-called "fast vertical migration pathways" between the interbedded aquiferous and weakly permeable rocks run deep. These pathways may primarily be associated with recent solid-rock tectonic fractures and unconsolidated rock disintegration. Changes in facies, mineralogical inhomogeneity, and granulometric inhomogeneity may also play a role here. The amount of openness in these pathways can only be determined using toxic indicators with very low detection thresholds—pesticides and radionuclides. No significant halos are observed for weakly toxic pollutants (sulfates, chlorine, various metals).

**5.4.2.3  Factors Influencing Groundwater Contamination.** The rate at which groundwater becomes contaminated with radionuclides depends on many factors. As Figures 5.4.2 and 5.4.3 indicate, the peak contamination for the subsurface water associated with the Eocene deposits occurs toward the center of the Kiev Metropolitan Area. At the center of the Eocene deposits, there is a depression in the subsurface water level from the Kiev water intakes which increase the recharge modulus of the multilayer aquifer system.

This correlation between subsurface water contamination and infiltrational recharge is confirmed by a high correlation coefficient (r = 0.98 for $^{137}$Cs and r = 0.88 for $^{90}$Sr in the Eocene aquifer).

The peak contamination for the subsurface water associated with the Quaternary deposits occurs to the north of the Kiev Metropolitan Area, in an area of high surface contamination.

Table 5.4.3 describes the correlations we have established between the radionuclide concentration in the subsurface water and the hydrogeological and radiochemical characteristics of the Kiev Metropolitan Area.

Further analysis of the results was performed using a multiple correlation method and expert data reliability assessment, enabling us to obtain a more stable estimate of the distribution of aquifer contamination.

Figure 5.4.4 shows the results of applying this method to the $^{137}$Cs contamination in the Eocene aquifer. We see that the basic surface distribution of the radionuclides remained unchanged, but the contours are smoother; in our view, this is a more accurate representation of the actual distribution of $^{137}$Cs in the subsurface water under the Kiev Metropolitan Area.

**5.4.2.4  Monitoring Needs.** In spite of the fact that contamination levels for much of the groundwater are currently low, the subsurface water monitoring effort should be improved. Also, full-scale in situ studies of radionuclide migration and redistribution should be conducted.

**TABLE 5.4.3.** Radionuclide Concentration in Subsurface Water in Eocene and Quaternary Deposits and Properties of the Overlying Contaminated Area

| Hydorgeological/Radio-chemical Properties of Area Under Study | Quaternary Aquifer | | Eocene Aquifer | |
|---|---|---|---|---|
| | $^{137}$Cs | $^{90}$Sr | $^{137}$Cs | $^{90}$Sr |
| | Correlation Coefficient | | | |
| Recharge | 0.97 | 0.95 | 0.98 | 0.88 |
| Depth of depression in water surface | | | 0.75 | 0.65 |
| Thickness of first confinement layer | | | -0.63 | -0.2 |
| —Humus content | -0.63 | -0.98 | | |
| —Clay content | -0.99 | -0.65 | | |
| Local surface contamination | 0.72 | 0.4 | | |
| $^{137}$Cs concentration in subsurface water, mBq/L | | 0.5 | | 0.7 |
| $^{90}$Sr concentration in subsurface water, mBq/L | 0.5 | | 0.7 | |

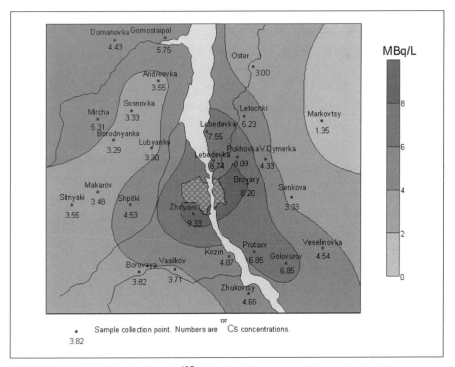

**FIGURE 5.4.4.** Distribution of $^{137}$Cs in the Eocene aquifer under the Kiev Metropolitan Area (including multiple correlation effects)

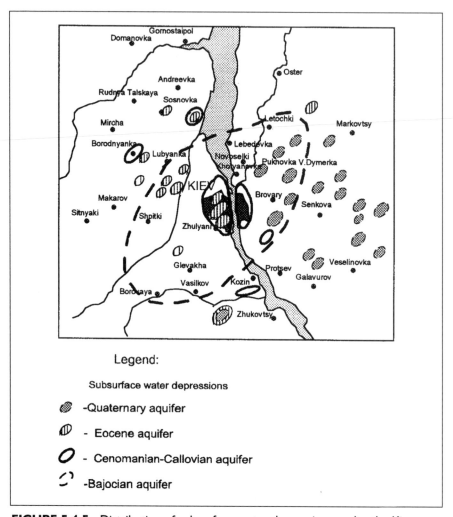

**FIGURE 5.4.5.**  Distribution of subsurface water depressions under the Kiev Metropolitan Area (including multiple correlation effects)

The first subject of this radiohydroenvironmental monitoring effort should involve monitoring the subsurface potable water resources of the Kiev Metropolitan Area.

Analysis of the current network of monitoring wells within the Kiev area shows that it is inadequate to monitor subsurface water levels. For example, there are only 19 monitoring wells within the city of Kiev and 8 wells in the remainder of the Kiev Metropolitan Area to monitor the aquifers currently in use. All of the wells for the Neogene and Paleogene aquifers are located within the city of Kiev (14 and 9 wells, respectively).

This clear lack of wells in the official network for monitoring the development of the subsurface water depression under the city of Kiev is partially compensated for through quarterly measurements of subsurface water levels (for the aquifers currently in use) in the 44 wells operated by Kievvodkanal. However, there are no wells to make up for the lack of monitoring at the edge of the depression.

Hydrochemical and radiological studies of subsurface water in the Kiev Metropolitan Area are performed by organizations under the Ukrainian State Committee for Geology, PO Kievvodkanal, and the Kiev City and Oblast Public Health and Epidemiological Services. Macrocomposition, synthetic surfactants, pesticides, microcomposition (via spectral analysis), and radionuclides are measured. Most of these analyses are based on determining whether or not the concentration is above (or below) the maximum permissible concentration. Thus, in the majority of cases, virtually no observations that would enable the measurement of variations in this parameter with time (seasonal trends, general trend, or other variations) for the subsurface water have been performed. In addition, the number of sampling points is limited. The PO Kievvodkanal and Kiev City and Oblast Public Health and Epidemiological Services laboratories primarily analyze water provided to users (frequently mixed surface and subsurface water), which prevents the characteristics of the subsurface water in its natural or artificially altered state from being determined.

Thus, only areas within the existing official network are usable in the radiogeological monitoring system currently under development for the Kiev Metropolitan Area, and then only after careful evaluation of each individual well. Without this comprehensive, reliable monitoring system, enough information will not exist to make decisions about groundwater.

## 5.5 CHARACTERISTICS AND MOBILITY OF RADIONUCLIDES

### 5.5.1 Behavior of Radionuclides in Soil

Analysis of experimental and literature data indicates that the radioactive fallout from the Chornobyl NPP accident came in a wide variety of physical and chemical forms. Because of the spatial inhomogeneity of the fallout on the Earth's surface, the variety of soil types within the contaminated area, and the variety of physical and chemical types of fallout present, the Chornobyl release was characterized by a wide variety of transformational dynamics in the soil. The transformational dynamics led to various rates of uptake and transport for geological and biochemical migration chains.

Using the basic laws governing the behavior of biologically significant radionuclides ($^{90}$Sr and $^{137}$Cs) in the soil and soil-vegetation burden, there are three major groups of parameters that determine radionuclide mobility in the soil and biochemical migration chains:

- Soil-migration chain: the concentration of stable radionuclide carriers in mobilized form, the acidity of the soil solution, the organic matter content, the mechanical and mineralogical composition of the soils, and the moisture content of the soils
- Terrain and geochemistry: local landscape and geochemical conditions (the location of the site in question relative to other terrain features), the origin of the topsoil and vegetative cover, and the type of plant growth (natural or artificial)
- Vegetation chain: physiological characteristics of the vegetation and the structure of the root system.

The radionuclides in the soil and vegetative cover as a result of the Chornobyl NPP accident show several distinct behavior patterns that deviate from standard expectations. These deviations are a result of the new physical and chemical types of radioactive materials contained in the Chornobyl fuel particles, as well as the characteristics of the topsoil and vegetative cover in contaminated areas of the Polessye.

Analysis of the experimental data revealed that the fuel component of the Chornobyl NPP release could have a substantial impact on known radionuclide uptake processes in various biological and geological migration chains.

5.5.1.1 **Condensation Fallout and Distant Plumes.** The migration of radionuclides in the condensation component of the fallout was basically well understood from pre-Chornobyl research. The contamination in the distant plumes was largely determined by the condensation component of the fallout. The physical and chemical characteristics of this fallout (especially solubility) are quite similar to those of global fallout.

Relatively accurate estimates for the mobility of the main radiologically significant radionuclides ($^{90}$Sr and $^{137}$Cs) can be obtained from known data describing the behavior of these radionuclides in global fallout on specific soils. Studies showed $^{90}$Sr and $^{137}$Cs are highly mobile in regions affected by condensation fallout and are thus available for uptake by the roots of plants [IAEA 1996a].

Once in the soil, cesium's biological availability roughly equaled that of cesium radioisotopes introduced into the same soils in water-soluble form (the relative biological availability of $^{137}$Cs fallout in soils from distant plumes in 1988 ranged from 0.9–1.0) [Bondar' et al. 1991]. The physical, chemical, and hydrological properties of the soils largely determined biological availability.

Because the condensation component of the fallout is more prominent at larger distances, the soil concentrations of mobilized forms of radionuclides will increase with distance from the NPP [Krivokhatskii 1990b; Kruglov et al. 1996]. The relative mobilized fraction of $^{90}$Sr ($^{90}$Sr:stable Sr) in the soil-water system increased with distance from the source [Salbu et al. 1996].

### 5.5.1.2 Condensed Fallout in the Inner Plumes.

Inner plumes can be characterized as a superposition of the condensation component and the fuel component of the fallout. The soil mobility of radionuclides in the inner plumes largely depends on the particle size and extent to which the particle matrix has undergone physical or chemical transformation.

Fundamental differences in behavior between the two main components of the fallout (condensation and fuel) were observed following the accident. By 1987, the soil concentrations of exchangeable $^{90}$Sr and $^{137}$Cs within the fuel-fallout plumes were a factor of 3–4 below those in the condensation plumes [A.N. Arkhipov 1995]. The smallest radionuclide leaching levels (a fraction of a percent of the concentration in the soil) were observed in areas where the density of fuel-particle fallout was highest [Krivokhatskii et al. 1990a; Kruglov and Arkhipov 1996]. The extent of radionuclide leaching due to atmospheric washout was studied in several samples of derno-podzolic sandy soil collected near the center of the 30-km Exclusion Zone in 1986. The values ranged from 0.26–0.36% of the radionuclide concentration in the sample for $^{95}$Zr, $^{95}$Nb, $^{134,137}$Cs, and $^{141,144}$Ce, and from 0.62–0.9% for $^{103,106}$Ru [Arkhipov et al. 1994]. During the initial years after the fallout, the concentration of exchangeable $^{90}$Sr in the soil within the fuel fallout plumes (1.5–12%) was smaller than that of $^{137}$Cs (1.0–38%).

Research indicates that extractable radionuclide concentrations (i.e., the extractable radionuclide fraction, which is less tightly bound to the soil solid phase) decrease with time. The rate of decrease depends on the initial physical and chemical characteristics of the radioactive fallout and the properties of the soil [Ivanov et al. 1991; Ivanov et al. 1992]. Generally speaking, the concentration of high-mobility $^{137}$Cs in soil in the distant fallout plumes was much higher than that in the inner plumes. This higher concentration was primarily because the condensation component of the fallout was dominant in the inner region [Salbu et al. 1994].

### 5.5.1.3 Mobility of $^{137}$Cs and $^{90}$Sr in Soil.

The presence of low-solubility fuel particles in the fallout is what led to the basic differences in the soil mechanics of high-mobility $^{137}$Cs and $^{90}$Sr. As shown by Konoplev et al. [1992], the rate constant for sorption of $^{137}$Cs by the soil solid phase is much greater than the rate constant for leaching of this radionuclide from the particle matrix; in most cases, this will lead to a decrease in the concentrations of the mobilized radionuclide forms as a function of time. For example, data obtained by Kruglov and Arkhipov [1996] suggest that the amount of water-soluble $^{137}Cs$ in the soil under the fuel fallout plume was extremely small—no greater than 1% in any of the years covered by the study. From 1987–1989, a gradual decrease in the concentration of exchangeable $^{137}Cs$ was observed, accompanied by a simultaneous increase in the acid-soluble fraction. Over the next several years, the acid-soluble fraction also began to decrease, and the percentage of tightly bound $^{137}Cs$ in the soil upon completion of all processing increased [Kruglov et al. 1996].

In the case of $^{90}Sr$, where ion exchange is the main mechanism for sorption on the soil solid phase and fixation processes are extremely slow, the pattern is quite different. As the fuel particles disintegrate, the soil concentration of mobilized $^{90}Sr$ increases at a rate directly proportional to the particle transformation rate. The estimated fuel-particle transformation rates obtained over the first 5–6 years following the accident were fairly inconclusive. Experts from the Russian Academy of Sciences Radium Institute found that by 1989 the fuel particles had begun to decompose, thereby causing the radionuclides to become mobilized [Krivokhatskii et al. 1990a]. According to some estimates, the concentration of mobilized $^{137}Cs$ peaked 1.5–2 years after the accident, and the total $^{137}Cs$ concentration peaked 6–10 years after the accident, depending on soil facies [Bondarenko 1994]. For $^{90}Sr$, the disintegration rate for the fuel particles and rate of transfer into the soil solution are higher than the fixation rate on solid soil components; this causes mobilization and subsequent increased biological availability of $^{90}Sr$ in areas contaminated by fuel particles [Salbu et al. 1996; Yakushev et al. 1996].

5.5.1.4 **Transformation Rates and Soil Mobility.** Destruction of oxidized fuel particles played a dominant role in radionuclide migration in soil during the first 2–4 years following the release  (e.g., the North and South Plumes) [Meshalkin et al. 1990, Skorobogatko and Rybalko 1992]. In areas where the fallout consisted of slightly oxidized fuel particles (the West Plume), the transformation processes are much less rapid, and 10–15 years will be required for the specific behavior of $^{90}Sr$ in the topsoil and vegetative cover to develop. The term "transformation" is used in several references [Ivanov and Kashparov 1992; Ivanov et al. 1996a] to mean any process that causes a radionuclide to be removed from the fuel-particle matrix (primarily particle destruction and dissolution). Table 5.5.1 lists some estimated transformation rates for fuel plume particles in soil under laboratory and natural conditions.

**TABLE 5.5.1.** In Situ and Model Estimates of the In-Soil Transformation Rates of Fuel Particles ($K_f$)

| Conditions and Soil | $K_f, s^{-1}$ | Source |
|---|---|---|
| Large fuel particles, laboratory conditions | $2 \times 10^{-11}$–$4 \times 10^{-11}$ | Skorobogatko and Rybalko 1992 |
| Natural conditions, soddy acidic alluvial soil | $2.0 \times 10^{-8}$ | Konoplev and Bulgakov 1992 |
| Natural conditions, derno-weakly podzolic sandy soil | $4.6 \times 10^{-9}$ | Konoplev and Bulgakov 1992 |
| Natural conditions, derno-weakly podzolic sandy soil (plowed fields) | $4.2 \times 10^{-8}$ | Konoplev and Bulgakov 1992 |
| Natural conditions, alluvial soils | $1.3 \times 10^{-9}$–$5.8 \times 10^{-9}$ | Demchuk et al. 1990 |
| Model simulations | $8.6 \times 10^{-9}$–$9.9 \times 10^{-8}$ | Borovoi 1990 |
| Natural conditions, using isotopic dilution method: | | Kashparov et al. 1996 |
| • Highly oxidized particles north and south of Chornobyl NPP: $V = 1.45 \times 10^{-7} \times 10^{-0.25xph*}$ ($s^{-1}$) | $1.45 \times 10^{-8}$–$2.6 \times 10^{-9}$ | |
| • Slightly oxidized particles in west plume: $V = 4.38 \times 10^{-7} \times 10^{-0.46xph*}$ ($s^{-1}$) | $6.3 \times 10^{-9}$–$2.6 \times 10^{-10}$ | |

*pH of water extract from soil.

Several experimental studies have shown that the dynamics of the fuel-particle transformation process in soil are determined by the origin of the particles, the particle size, and the soil chemistry. The spatial distribution of particles whose matrices have undergone various levels of physical and chemical transformation are highly nonuniform.

### 5.5.2 Dynamics of In Situ Soil Studies of Radionuclide Transformations

Experimental studies of the physical and chemical forms and processes involving [137]Cs and [90]Sr in the soil under various release plumes were conducted. Soil samples with various physical, chemical, and mechanical properties were used. In addition, substantial differences in the initial physical and chemical properties of the fallout were included. The concentrations of various mobilized radionuclide forms in the soil were determined by sequential extraction [Pavlotskaya 1974].

The studies showed that the exchangeable[137]Cs soil concentration generally decreases with time. The rate of decrease varied because of variations in soil chemistry as well as differences in the physical and chemical properties of the fallout. The exchangeable[137]Cs concentration in the soil (given soils with similar properties under similar landscape and geochemical conditions) generally increases with distance from the point of release. This trend is due to spatial variations in the physical and chemical properties of the fallout.

Hydromorphic soils, which typically have a higher cation-exchange capacity and a higher clay-mineral content than automorphic soils, also have much lower exchangeable cesium concentrations and larger decreases in the exchangeable cesium concentration as a function of time [Sanzharova et al. 1994]. However, there are several exceptions to this trend. Both meadow soil and arable soil in the 30-km Exclusion Zone showed high concentrations of exchangeable [137]Cs until 1990–1991, at which point the exchangeable [137]Cs concentrations decreased. The low levels of exchangeable radionuclides in the inner region near Chornobyl NPP (5–10 km inner Exclusion Zone) is due to the soil properties, as well as the fact that the fuel component contributes a large fraction of the total soil contamination. Experimental data obtained by isotope dilution [Ivanov et al. 1992] showed rapid radiocesium sorption by the solid phase of hydromorphic soils when the radiocesium was initially added to the soil in water-soluble form. The sorption was faster on hydromorphic soils and automorphic soils. This result is consistent with the lower concentrations of exchangeable radionuclides for the first 2 years after the radionuclide was introduced into the soil (0.5–5% as opposed to 15–30% in automorphic soils).

### 5.5.3 Physical and Chemical Transformation of Radionuclides in Agricultural Soils and Countermeasures

Four countermeasures were used to reduce the soil-to-vegetation transfer coefficients for cesium radionuclides in agricultural soils that were contaminated by the Chornobyl NPP accident. These countermeasures are

- Soil amendments with high cation-exchange capacities (zeolite, clinoptilolite, etc.)

- Soil amendments that sorb cesium radioisotopes via mechanisms other than ion exchange (hydromicas, montmorillonite, etc.)

- Chemical amendments to change soil chemistry, in particular, the pH of the soil solution (lime, ground dolomite, etc.)

- Mineral and organic fertilizers (manure, increased doses of potassium fertilizers, and potassium and phosphorus fertilizers).

The dose effectiveness of agrochemical measures largely depends on the extent to which they target specific components of the agricultural ecosystem (e.g., the soil layer, plant metabolism). The first three measures targeted the soil layer and enhanced mechanisms (ion-exchange sorption, fixation, coprecipitation, etc.) that bind radionuclides to the soil solid phase and reduce the concentration of radionuclides in the soil solution. With the cesium bound to the soil, the transfer of cesium to vegetation is limited. Given the specialized nature of the soil cover in contaminated areas of the Polessye (various types of light-textured derno-podzolic soil and peat-bog soil), the agrochemical countermeasures implemented were targeted at the soil layer.

The fourth measure, mineral and organic fertilizers, affects both the soil layer (increase the concentration of stable radionuclide carriers and/or the concentration of organic matter in the soil solid phase and the soil solution) and the vegetation (physiological) layer. The fertilizers affect the plant metabolism, including mineral exchange, nutrient uptake from the soil, and increased plant productivity, etc. The research results shown in Figure 5.5.1 indicate that when mineral and organic fertilizers are added to the soil, this reduces the soil-to-vegetation transfer coefficients for oats (grain and straw), lupine (pods), and *Brassica napus*[1] (also known as rape) (vegetation) by approximately a factor of 2 [Bondar' et al. 1996]. However, these countermeasures do not significantly change the soil concentrations of mobilized cesium radioisotopes, as noted by the successive extraction scheme being used.

Experimental research methods for studying 1) the physical and chemical forms of radionuclides in soil and 2) the transformations resulting from the application of mineral and organic fertilizers may be useful tools for assessing the role of soil and vegetation layers in reducing the soil-to-vegetation transfer coefficient. By better understanding the transfer coefficient, more effective countermeasures could be developed and implemented to limit the soil-to-vegetation transfer of radionuclides.

### 5.5.4 Soil Properties and Mobilized Radionuclide Concentrations

The potential relationships between extractable $^{137}$Cs soil concentrations and soil properties were identified by structural analysis using the Spearman rank correlation coefficient. Any potential effects due to type of fallout were minimized by only using extractable radionuclide data obtained in

---

[1]Editor's Note: *Brassica napus* is popularly called canola in the United States.

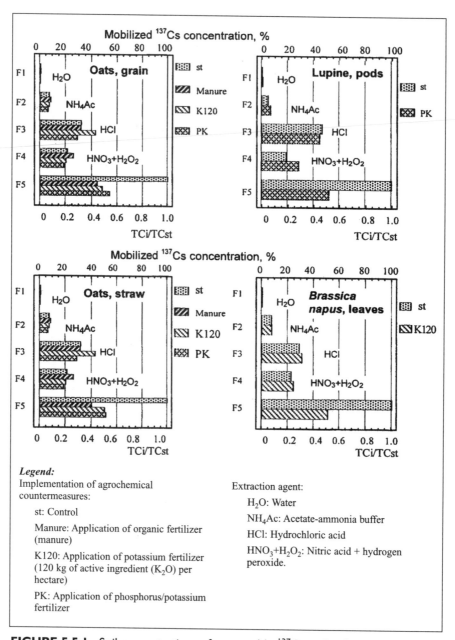

**FIGURE 5.5.1.** Soil concentrations of extractable $^{137}$Cs and soil-to-vegetation transfer coefficients after implementing various agrochemical countermeasures

1991–1994. Note, the variations in the concentrations of specific extractable radionuclides from one sample plot to another are much larger than the variations within the soil of a single sample plot from 1991–1994. The soil parameters involved in the analysis are described in Table 5.5.2.

**TABLE 5.5.2.** Soil Parameters Analyzed

| Parameters | Mineral Soils | Biogenic Soils |
|---|---|---|
| $pH_{KCl}^{(a)}$ | 3.9–7.3 | 3.8–7.5 |
| $C_{org}^{(b)}$, % | 0.3–10.1 | 7.7–57.1 |
| Cation exchange capacity, meq/100 g | 1.2–60 | 7.3–72 |
| $K_{mob}^{(c)}$, mg/100 g | 0.1–42 | 1.1–94 |
| Alphitite concentration, % | 0–55.2 | |

*References:* Belli et al. 1994; Bunzul et al. 1994; Sanzharova et al. 1994; Prister et al. 1996a; Ivanov et al. 1996a.
(a) $pH_{KCl}$ is the salt pH (concentration of hydrogen ions in the salt extract obtained upon extraction with 0.1N KCl)
(b) $C_{org}$ is the total organic carbon
(c) $K_{mob}$ is the mobilized potassium concentration

Sequential extraction of the soil samples showed that the radiocesium was distributed over the extractable fractions as follows (% of total soil concentration):

- Readily exchangeable: 0.5–17.3

- Iron oxide-bound and manganese-oxide bound: 0.2–12.6

- Organic-matter bound: 0.8–52.5

- Tightly bound: 12.6–61.0

- Nonextractable residue: 10.0–78.7.

For organogenous soils, the extractable fractions are as follows (percent of total soil concentration):

- Readily exchangeable: 0.3–2.2

- Iron oxide-bound and manganese-oxide bound: 0.4–1.7

- Organic-matter bound: 4.4–16.6

- Tightly bound: 16.5–33.7

- Nonextractable residue: 50.7–70.5.

Thus, most of the $^{137}Cs$ in the soils is contained in the tightly bound and fixed fractions.

The data indicate that the  much higher in high-tilth soils characterized by high concentrations of clay minerals and high  of the organic-bound fraction and $pH_{KCl}$ (for organogenous soils), and the organic-matter content, cation-exchange capacity, and exchangeable potassium content (for organogenous and mineral soils). Note, the  correlations in the organogenous soils may be distorted by the small sample size ($N = 7$).

Correlation analysis of extractable $^{137}Cs$ concentrations determined using the Pavlotskaya extraction method showed moderate correlations ($r = 0.575$–$0.722$) between the water-soluble radionuclide concentration and the exchangeable potassium content (for organogenous soils), $pH_{KCl}$ and the organic-matter content, cation exchange capacity, and exchangeable potassium content (for mineral soils).

Moderate correlations ($r = 0.406$–$0.427$) were observed between the ion-exchangeable $^{137}Cs$ concentration and the organic-matter content, cation-exchange capacity, and exchangeable potassium content (for mineral soils). Similar correlation levels ($r = 0.470$–$0.683$) were observed between the mobilized $^{137}Cs$ concentration (Pavlotskaya, for example, extracted with 1N HCl solution) in mineral soils and soil properties such as clay content, organic-matter content, cation-exchange capacity, and exchangeable calcium content.

Fairly strong correlations ($r = 0.702$–$0.882$) were observed between the concentrations of ion-exchangeable $^{90}Sr$ and soil properties such as clay content, cation-exchange capacity, and exchangeable calcium content.

Thus, the concentrations of the radionuclide phases (fractions) that are less tightly bound to soil solid phase components decrease with time. The rate of decrease depends on the initial physical and chemical properties of the radioactive fallout, as well as the soil properties. Generally speaking, the mobile $^{137}Cs$ soil concentrations are much higher in the outer release plume than in the inner plume; this is primarily due to the fact that the condensation fallout component predominates in the outer region. Thus, data on the extractable $^{137}Cs$ concentration can be used as a criterion for measuring the radionuclide dynamics in soil.

Nevertheless, the lower mobilized $^{137}Cs$ concentrations in the inner region are not solely due to the fact that the fuel component makes a larger contribution to the total surface contamination than in the outer region, but also to the soil properties. Thus, obtaining adequate predictions of $^{137}Cs$ redistribution in various components of the soil and vegetation layer and the planning and implementing countermeasures will require that environmental characteristics (such as terrain and geochemistry) be taken into account in addition to anthropogenic factors (physical and chemical properties of the fallout, etc.).

In mineral and organogenic soils, comparing experimental data on the mobilized $^{137}$Cs concentration against the soil-to-vegetation transfer coefficients suggests that the mobilized $^{137}$Cs concentration cannot be directly used as a indicator of the biological availability of the radionuclide to vegetation.

Studying the dynamics of the extractable cesium radioisotopes along various Chornobyl NPP release plumes showed a general decline in the soil concentration of mobilized radionuclides as a function of time. However, this trend did not apply to all of the experimental plots in the 30-km Exclusion Zone and is not valid under all soil conditions or contamination characteristics. In particular, some of the experiments showed that the exchangeable $^{137}$Cs concentrations increased from 1988–1991 for organogenous soil primarily contaminated by the fuel component of the fallout.

Research studies performed by the Belarus Scientific Research Institute for Radiology indicated that the relative concentration of exchangeable radionuclides in virgin soil varies from 1.5–3%, while that in plowed soil ranges from 5–10%. Elevated $^{137}$Cs concentrations relative to virgin soil have also been observed in meadow soil following grass regeneration and application of soil amendments [Firsakova et al. 1990]. The concentration of exchangeable and acid-soluble $^{137}$Cs was consistently higher in plowed soil than virgin soil over the period of time covered by the research (1988–1991).

Experimental results indicated that the concentration of $^{137}$Cs in the solid residue after sequential extraction of virgin soil samples was much lower than that for plowed soil, even though the concentration of exchangeable radionuclides in such soils was lower or virtually identical to that in plowed soils. The concentration of mobilized (water soluble or ion-exchangeable) and acid-soluble radionuclides in the plowed soil was shown to gradually decrease over the period of time covered by the studies (1988–1991), with a corresponding increase in the concentration of radionuclides in the soil-sample residue remaining after sequential extraction. Thus, agricultural utilization and tilling lead to more rapid fixation of $^{137}$Cs to the soil solid phase. Cesium is more tightly bound in organic soils that have been plowed and planted than undisturbed soil.

### 5.5.5 **Vertical Transport of Radionuclides in Soil**

To understand and assess radionuclide migration in soil and biological chains, studies were performed on the physical and chemical transformations of radionuclides and the rate of vertical radionuclide redistribution in soil for the Chornobyl NPP plumes taking into account the properties of the soil and fallout. Studies of each plume performed in the first few years following the accident were inconclusive. However, several trends were noticeable:

- The transfer rate for vertical redistribution of the radionuclides in the release was quite slow [Bondar' et al. 1989; Novikova 1989; Petryaev et al. 1989]

- There was no difference in the type of migration or migration rate for radioisotopes of different chemical elements (cesium, ruthenium, cerium, and strontium) [Bondar' et al. 1989]

- There was little spatial differentiation in the radionuclide concentrations ($^{134,137}$Cs and $^{90}$Sr) for various terrain components [Karavaeva et al. 1990].

These trends may be due to the highly patchy nature of the soil cover in the contaminated area, as well as the relatively inhomogeneous distributions of individual radionuclides within the fallout, as well as the physical and chemical properties of the fallout.

Conducting full-scale experiments and observations to identify and analyze the major behavior patterns of the biologically important radionuclides ($^{137}$Cs, $^{90}$Sr, and $^{239,240}$Pu) in the soil throughout the contaminated area would be extremely difficult for several reasons. These reasons are as follows: the spatial structure of the contamination is highly nonuniform, the fallout has a wide range of physical and chemical properties, and the soil cover is nonuniform. Thus, laboratory simulations supplemented full-scale experiments. Laboratory simulations enable the number of factors affecting radionuclide transport in soil to be reduced, and enable assessment of the significance of individual migration processes (mechanisms) to the overall vertical radionuclide redistribution process.

Analysis of the experimental data shows that the vertical rate of transport for $^{137}$Cs in the soil is highly dependent on the nature of the soil cover (e.g., the physical, chemical, and hydrological characteristics of the soil). In particular, automorphic mineral soils were observed to have the smallest migration coefficients ($3.0 \times 10^{-9}$–$5.0 \times 10^{-9}$ cm$^2$/s), while hydromorphic organogenous soils were observed to have much larger values—at least $1.0 \times 10^{-8}$–$3.0 \times 10^{-8}$ cm$^2$/s [Ivanov et al. 1996b].

The experimental data show slight variation in the $^{137}$Cs transport parameters for automorphic soils, even when the properties of the soil and vegetation layer (surface roughness, sod thickness, height of vegetation, moisture content of soil, etc.) are relatively uniform. However, substantial variations (a factor of 3–6) are observed in the vertical radionuclide transport parameters for soil and vegetative cover with the following characteristics: derno-podzolic sandy soil in plots with normal moisture levels; the same soils with various degrees of gleyization in plots with higher moisture levels; and peat and gley soils in swampy areas. Data on the vertical $^{137}Cs$ distribution indicate that the migration mobility for $^{137}Cs$ is determined by soil moisture as well as soil type [Ivanov et al. 1996a; Ivanov et al. 1996b] (Figure 5.5.2).

Analysis of the experimental data indicates that between 1989 and 1996, the $^{137}Cs$ vertical transport parameters varied only slightly (or by less than the measurement error) from one experimental plot to another. Because the vertical transport parameters varied only slightly, we can use the $^{137}Cs$ transport parameter in various soils to predict radionuclide migration using generally accepted transport models (quasidiffusion or convective-diffusion models).

Experimental data for the vertical distribution of $^{90}Sr$ and plutonium isotopes in the experimental plots and the transport coefficients for these radionuclides calculated using a convective diffusion transport model indicate that $^{90}Sr$ transport in the soil is more rapid than $^{137}Cs$ or plutonium transport. However, the vertical transport coefficients for $^{90}Sr$ and plutonium are similar in plots where the fuel component of the fallout makes up a high

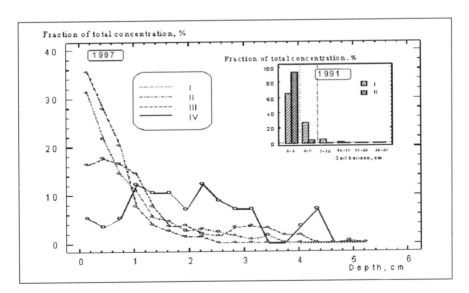

**FIGURE 5.5.2.** Vertical distribution of $^{137}Cs$ at various points in the geochemical profile in 1987 and 1991 (Pripyat River floodplain, R = 37 km; NW; August 1987): I—eluvial terrain (drainage divide); II—transeluvial terrain; III—superaqueous terrain ("saucer"); IV—superaqueous terrain (periodically flooded area)

percentage of the total fallout; this appears to be the result of the dominant role played by mechanical fuel-particle transport in the vertical direction. The vertical $^{90}$Sr transport rate for such plots is much lower than for plots where the fuel component makes a smaller contribution to the total contamination [Ivanov et al. 1996a; Ivanov et al. 1996b].

During the first few years following the 1986 accident virtually no differences were observed between the vertical distributions for cesium, cerium, ruthenium, and strontium, which have different physical and chemical properties. The vertical distributions of these four elements were later observed to become differentiated [Ivanov et al. 1996b].

Studies of the behavior of global fallout radionuclides in various components of the soil and vegetative cover showed that site terrain and geochemistry had a substantial effect on the dynamics of $^{90}$Sr and $^{137}$Cs redistribution within each of these components [Pavlotskaya 1974; Prister et al. 1991]. The nature of radionuclide redistribution within terrain elements is highly dependent on geomorphology—direction of runoff, sod coverage, aspect, shape of a depression, etc. [Prister et al. 1991]. On the one hand, under certain conditions, radionuclides are observed to accumulate in enclosed depressions with plowed slopes [Pavlotskaya 1974; Prister et al. 1991]. On the other hand, the $^{90}$Sr concentration was observed to be much lower in small flat-bottomed valleys within the steppe zone than on drainage divides.

The Ukraine Polessye is characterized by relatively large variations in terrain and geochemistry. Swampy soils, derno-podzolic gley soils, peat-gley soils, and weakly and moderately podzolized soils underlain by sand predominate in the area surrounding the headwaters of the Pripyat River and on the fluvial terraces above the floodplain over the entire length of the river. Groundwater lies at a depth of 1.5 m and frequently comes to the soil surface. The high moisture content causes rapid leaching of derno-podzolic sandy soils [Vil'gusevich 1955; Lukashev and Petukhova 1965; Petukhova 1986].

Additional terraces, located dozens of kilometers (in certain areas, 100–150 km) from the Pripyat floodplain, are dominated by various derno-podzolic soils (sand, loamy sand, and loamy soils). The depth to groundwater gradually increases from the floodplains of the Pripyat River and its tributaries outward to the boundaries of the watershed, varying from 0.5–3 m [Lukashev 1961; Kulakovskaya 1965].

Data on the vertical redistribution of radionuclides associated with various landscape elements were obtained at various points within a plot in the Pripyat floodplain (the Osinovitsa tract, near Khvashchevka, Khoiniki Region, Belarus). The vertical $^{137}$Cs distributions (Figure 5.5.2) indicate that the highest radionuclide transport rates are observed in the plots with high moisture content (Points III and IV, which correspond to superaqueous terrain) relative to dry meadow plots on derno-podzolic sandy soil (Points I and II, which correspond to eluvial and transeluvial terrain).

### 5.5.6 Basic Factors Affecting the Vertical Transport Rate for Radionuclides

Cerium, plutonium, strontium, and other refractory elements, which fell as highly active nuclear fuel particles, are more likely not to follow the global fallout vertical distribution behavior. Radionuclides such as $^{137}$Cs that make up the condensation fallout are more likely to follow the global fallout vertical distribution. The radionuclide $^{137}$Cs has similar migration mobility and biological accessibility to global fallout.

Generally, $^{90}$Sr has higher mobility than $^{137}$Cs or the plutonium isotopes for various types of soil characterized by ratios of fallout types. This is because these soils have a lower sorption capacity for $^{90}$Sr and the specific nature of the processes that determine the phases of this radionuclide present in the soil (most of the $^{90}$Sr is in phases that are sorbed through ion exchange).

Simulation data show elevated diffusive transport rates for $^{90}$Sr and $^{137}$Cs in various types of soil (derno-podzolic sand, derno-podzolic sandy loam, and peat-bog soil) at high moisture content. The largest increase in transport rate associated with high moisture content for both $^{137}$Cs and $^{90}$Sr was observed in peat-bog soil. As noted previously, this phenomenon is primarily due to the fact that the increase in absolute moisture content due to elevated moisture content relative to saturation was higher in peat-bog soils than in mineral soils (increasing the moisture content from 30–100% of saturation corresponds to an increase in absolute moisture content from 9–30 mass% for mineral soils and 33–110 mass% for peat-bog soil). This elevated moisture content leads to an increase in the fraction of radionuclides transported into the soil profile as a result of diffusion in the pore solution and directional transport flowing in the same direction as the soil moisture [Ivanov et al. 1996a].

The simulations also showed that radionuclide properties cause $^{137}$Cs, $^{90}$Sr, and $^{239}$Pu to have different migrational mobilities in the same soil. Cesium-137 is most mobile in peat-bog soil (diffusion coefficient $D = 1.7$ x $10^{-8}$ cm$^2$/s for a moisture content equal to 30% of saturation) and least mobile in derno-podzolic sandy soil (diffusion coefficient $D = 0.5$ x $10^{-8}$ cm$^2$/s for a moisture content equal to 30% of total moisture content). Strontium-90 has the highest diffusive transport rate in derno-podzolic sandy soil ($D = 20.0$ x $10^{-8}$ cm$^2$/s at a moisture content equal to 30% of saturation) and the lowest in peat-bog soil ($D = 0.6 \times 10^{-8}$ cm$^2$/s at a moisture content equal to 30% of saturation) [Ivanov et al. 1996a].

Analysis of experimental data indicates that there is little variation in the $^{137}$Cs distribution for all automorphic soils having similar physical properties, chemical properties, and mechanical composition. In 1995, approximately 80–95% of the $^{137}$Cs in the experimental plots was in the 0- to 5-cm layer.

Comparison of the $^{137}$Cs vertical distributions in release plume plots characterized by different ratios of the fuel and condensation components did not show any substantial differences. Because of the high sorption rate of $^{137}$Cs in mineral soils, most of the $^{137}$Cs is sorbed, and therefore much less involved in the vertical migration process (relative to the mobile forms of the radionuclide). The sorption rate for this radionuclide is much higher than the rate at which it is leached from the fuel-particle matrix [Konoplev and Golubenkov 1991]. The leached radionuclide transitions almost immediately to a sorbed state. Thus, the fuel component and sorbed $^{137}$Cs are fairly similar with respect to mobility (the migration rate for radionuclides in finely dispersed fuel particles is extremely low).

Results obtained within the framework of the modified transport model indicate that the peak migration rate for $^{137}$Cs and plutonium isotopes in automorphic soils occurred during the first 5 years following the accident, when the mobile component of these radionuclides played the most important role. The transport rate is currently determined by the fraction of the radionuclide that is in mobile form—a few percent of the total radionuclide content integrated over the soil profile—and tends to decrease with time. The decreasing percentage of radionuclides in mobile form should cause the biological accessibility to decrease with time.

The rate of radionuclide migration also depends on the fuel-particle disintegration rate. This dependence is especially strong for $^{90}$Sr, since the sorption rate for this radionuclide is an order of magnitude smaller than that of $^{137}$Cs or any of the plutonium isotopes.

Depending on soil chemistry and the characteristics of the fallout over the initial 5–15 years, leaching from fuel-particle matrices and sorption by soil minerals will remain the dominant processes that determine the rate of transport for $^{90}$Sr and plutonium isotopes in soil. The experimental material suggests that organogenous hydromorphic Polessye soils are the most critical from the point of view of migration and biological accessibility. For example, the highest observed migration rate for $^{90}$Sr occurred in periodically flooded peaty-podzolic soil plot. In this experimental plot, approximately 30% of the total radionuclides initially present had been removed from the top 4 cm of the soil profile by 1994. The radionuclide concentration was observed to peak in the 2–4 cm layer, which indicates that directed radionuclide transport with the flow of moisture made a substantial contribution to the vertical migration process.

The experimental data and the models of vertical radionuclide migration in the soil were used for preliminary long-term predictions of the concentrations of various radionuclide forms in the root zone. The effective times required for clearance of half the $^{137}$Cs from the 5-cm horizon of automorphic mineral soil in natural meadows will be 20–25 years; the corresponding value for hydromorphic organogenous soil in meadows will be 8–12 years. The effective times required for clearance of half the $^{90}$Sr from the 5-cm horizon of fine-grained automorphic mineral soil (sand) in natural meadows will be 6–8 years. The corresponding value for coarser-grained soil (loam) will be 10–15 years; while the value for organogenous soil will be 22–24 years. The effective times required for clearance of half the $^{90}$Sr from the plowed horizon on farmland will be 22–25 years for organogenous soil; 12–25 years for medium- and coarse-grained mineral soil; and 2–12 years for fine-grained soil. The effective times required for clearance of half the $^{90}$Sr from the plowed horizon on farmland will be 15–20 years for organogenous soil and 20–25 years for mineral soil.

Studies of the rate of fuel-particle dissolution and the rate at which radionuclides are converted into mobile phases suggest that the $^{90}$Sr contamination in agricultural products will peak 10 years after the accident in regions immediately to the south of the 30-km Exclusion Zone (Strakholissya, Gubin, Gornostaypil, etc.), and 20 years after the accident in the narrow south plume in Kiyiv and Zhitomir Oblasts (toward Dovhi Lis).

## 5.6 CONTAMINATION IN AGRICULTURAL PRODUCTS

The Chornobyl NPP accident contaminated approximately 8.4 million hectare of arable farmland in Ukraine. Countermeasures were and are needed to deal with the spread of this contaminated food. Countermeasures for a nuclear accident can be divided into three phases: first (early), second (intermediate), and third (late). In the first phase also known as the plume exposure phase, the primary source of radiation dose to the public is from the airborne plume of radioactive material coming from a facility. The routes of exposure include direct radiation from beta/gamma-emitting radionuclides in the plume (e.g., noble gases) and inhalation of radioactive material (e.g., radioiodine). Countermeasures include evacuation, shelter, and radioprotective drugs. In the intermediate phase also known as the ingestion exposure phase, the primary route of exposure is drinking from contaminated surface water sources, the iodine-grain-cow-milk pathway, ingestion of contaminated leafy vegetables, and some dose from deposition of beta/gamma emitters on the ground. Countermeasures include placing restrictions on water sources; using stored foods; placing cattle on stored feed; interdicting crops; and diverting dairy production to butter, cheese, or evaporated milk to allow the decay of $^{131}$I. In the late phase, the primary route of exposure is direct radiation from contaminated soil and the uptake of radioactive material into crops. Countermeasures include removing contaminated soil and monitoring agricultural products.

### 5.6.1 First Phase of Monitoring Agricultural Products

Shortly after the accident, contamination nationwide was determined using an aerial gamma-ray spectrometric mapping method with a sparse network of ground-based sampling. The data collected was not sufficiently detailed to deal with agricultural products, as the minimum sample spacing was generally at least 2–3 km.

There was no standard methodology for determining the contamination density on agricultural land. An analysis of mean regional cesium and strontium contamination densities was performed in June 1986 to locate fields with abnormally high contamination densities. Twenty-five samples, or five samples per field, were collected from five fields at each collective farm and state farm, and combined into a single regional sample. The concentrations of $^{134,137}$Cs and $^{90}$Sr were then determined at 11 Ukrainian laboratories. A map of the radionuclide concentrations on Ukrainian arable land was produced in late June 1986, and regions were identified for implementation of countermeasures or additional research [Ministry for Chornobyl Affairs 1996] (Table 5.6.1).

The accident contaminated more than 8.4 million hectare of farmland. In the six oblasts with the highest contamination levels (Volynska, Zhitomir, Kiev, Rivne, Chernigov, and Cherkasy), contamination of 3.7–37 kBq/m$^2$ was found on farmland. The percentage of farmland with 3.7–37 kBq/m$^2$ ranged from 52.3% for the Rivne Oblast farmland to 95.5% for Chernigov Oblast farmland. Contamination levels from 37–185 kBq/m$^2$ were found on farmland, the percentage of farmland contaminated ranged from 3.74% for the Chernigov Oblast to 44% for the Rivne Oblast. The maximum percentage of farmland with contamination densities of 185–555 kBq/m$^2$ was 4%. The Rivne and Volynska Oblasts farmland, 18.3% and 4.2% respectively, consists of peat bogs with contamination densities of 37–555 kBq/m$^2$.

Between 1990 and 1993, extensive measurements of contamination in milk and potatoes from private farms were performed. These results were used to categorize the dose burden for population centers in accordance with Ukrainian law. Internal exposure contributed 60% or more of the annual total effective equivalent dose in 80% of the population centers;

**TABLE 5.6.1.** Area of $^{137}$Cs-Contaminated Farmland, hectare in thousands

| Oblast | Area of Farmland Surveyed | Contamination Density, kBq/m$^2$ | | | | | | |
|---|---|---|---|---|---|---|---|---|
| | | 3.7–37 | 37–185 | 37–185 Peat Soils | 185–555 | 185–555 Peat Soils | >555 | Exclusion Zone |
| Vinnitsya | 936.1 | 850.0 | 85.7 | — | 0.4 | — | — | — |
| Volynska | 244.0 | 228.4 | 15.4 | 10.2 | 0.2 | — | — | — |
| Zhitomir | 1469.3 | 1116.8 | 270.3 | 34.2 | 62.0 | 9.3 | 20.2 | 0.7 |
| Ivano-Frankivsk | 91.4 | 71.3 | 19.1 | — | 1.0 | — | — | — |
| Kiyiv | 1528.8 | 1272.4 | 213.3 | 4.5 | 28.2 | 1.5 | 14.9 | 54.2 |
| Rivne | 329.3 | 172.1 | 145.7 | 48.8 | 11.5 | 4.1 | — | — |
| Sumy | 313.8 | 301.3 | 12.2 | 0.6 | 0.3 | 0.1 | — | — |
| Ternopil | 231.8 | 219.3 | 12.5 | — | — | — | — | — |
| Cherkasy | 1325.9 | 1172.8 | 146.6 | 1.2 | 6.5 | — | 0.05 | — |
| Chernivtsi | 97.5 | 74.5 | 22.7 | — | 0.3 | — | — | — |
| Chernihiv | 1833.7 | 1759.2 | 68.6 | — | 5.4 | — | 0.5 | — |
| Total | 8401.6 | 7238.1 | 1012.1 | 99.5 | 115.8 | 15.0 | 35.6 | 54.9 |

the percentage of population centers where this was the case was approximately 100% in Volynska Oblast, 99% in Rivne Oblast, 73% in Chernihiv Oblast, 70% in Zhitomir Oblast, 49% in Kiev Oblast, and 40% in Sumy Oblast. This type of analysis provides more substantial background for use in planning countermeasures and undertaking efforts to optimize agricultural production in areas where such methods will be effective [Ministry for Chornobyl Affairs 1996].

Direct prohibition on consumption of crops was especially effective for small individual farms, where most products are used fresh, without any processing. The urban population consumes milk and other products whose distribution is to a great extent centralized, which enables radionuclide intake to be restricted in an organized manner. In Kiev, for example, milk contaminated with $^{131}$I was diverted into the production of butter, which was then stored until the $^{131}$I concentration decreased below the maximum permissible concentration [Prister et al. 1993].

### 5.6.2 Soil pH and Radionuclide Uptake

Planning of countermeasures during the second phase requires monitoring; the targets of the monitoring should change as the situation develops and more accurate information becomes available. During this second phase, radiation monitoring should be supplemented by environmental monitoring to enable identification of critical regions and entities. For example, radiation and environmental monitoring showed that the Ukrainian Polessye was a critical area. The area has a wide range of geochemical properties. Mineral soils cover the spectrum from leached chernozem with a salt-extract pH varying from 6.6–7.5 to acid derno-podzolic soils with pH varying from 4.5–5.5. By understanding the amount of radiation in the area (radiation monitoring) and how the acidity of the soil influences the uptake of $^{137}Cs$ in fodder, the $^{137}Cs$ soil-to-feed transfer coefficient—2.5–13—was calculated and the variations in radionuclide uptake between different crop species—varies by a factor of 30–40—were determined (Table 5.6.2).

Variations in soil and climatic conditions occur at virtually every step in the food chain, so that significant variations in milk and meat contamination (milk and meat contribute approximately 70–90% of the human intake of $^{137}Cs$) are observed between different areas of the contaminated region.

With passage of time, more of the $^{137}Cs$ becomes fixed in the chernozem, which has a high capacity for absorption and retention of clay minerals As a result of cesium fixation in the chernozem, the soil-to-milk and soil-to-meat transfer coefficients for $^{137}Cs$ decreased by a factor of 4–5

**TABLE 5.6.2.** Soil-to-Fodder Transfer Coefficients for $^{137}Cs$, $(Bq/kg)/(kBq/m^2)$ (Mean for 1987–1990)

| | Soil type, ph of Salt Extract | | | |
|---|---|---|---|---|
| Crop | Derno-podzolic, 4.5–5.5 | Gray Forest 5.6–6.5 | Chernozems, 6.6–7.2 | Difference Times Smallest Value |
| Hay from natural grasses | 10.00 | 4.00 | 1.80 | 5.5 |
| Hay from planted grasses | 4.00 | 3.00 | 1.60 | 2.5 |
| Vetch | 2.70 | 0.45 | 0.20 | 13.0 |
| Clover | 1.80 | 0.30 | 0.30 | 6.0 |
| Lupine | 1.50 | 0.40 | 0.15 | 10.0 |
| Alfalfa | 0.80 | 0.40 | 0.20 | 4.0 |
| Silage | 0.40 | 0.20 | 0.08 | 5.0 |
| Fodder beet | 0.50 | 0.35 | 0.20 | 2.5 |
| Potato | 0.25 | 0.13 | 0.045 | 5.5 |
| Winter grain | 0.50 | 0.20 | 0.05 | 10.0 |
| Rye | 0.40 | 0.10 | 0.04 | 10.0 |
| Barley | 0.30 | 0.10 | 0.06 | 5.0 |
| Difference times smallest value | 40 | 40 | 30 | |

between 1987 and 1993, while the transfer coefficients for the acidic soils only decreased by a factor of 1.5–2 [Prister et al. 1993].

The peat soils, peat-gley soils, and peat bog soils which are widely distributed throughout the Polessye, have high organic-matter content (20–60%), an extremely low clay mineral and silt fraction, and an acid soil solution reaction ($pH_{KCl}$ 4.2–5.4). Such soils have soil-to-vegetation transfer coefficients of 3.7–30 (Bq/kg)/(kBq/m$^2$), while the derno-podzolic soils have transfer coefficients of 0.2–7.6 (Bq/kg)/(kBq/m$^2$) depending on the growing technique.

Radioenvironmental monitoring indicated that the following natural and seminatural ecological systems in the contaminated area were critical landscapes: forest land and natural or cultivated meadow land. Under Polessye conditions, a significant fraction of the meadow land consists of meadows on peat or peat-bog soil with various levels of gleization. Such soils have much higher soil-to-meadow vegetation transfer coefficients than arable soil (Table 5.6.3). As the data in Table 5.6.3 indicate, $^{137}$Cs accumulation in meadow grasses is a strong function of the water content of the soils and is much higher in low-lying pastureland or pastureland located in floodplains [Prister et al. 1993].

These soil characteristics imply that a radiation dose of 1 mSv due to external and internal exposure will be reached in areas where peat soils are common when the $^{137}$Cs contamination density decreases to 11–22 kBq/m$^2$ rather than 220–300 kBq/m$^2$, as in areas where chernozem soil is common (Table 5.6.4).

**TABLE 5.6.3.** $^{137}$Cs Content in Sod and Herbage for Soil Contamination Densities of 1 kBq/m$^2$ in Natural Meadows (1988–1989)

| Soil | Meadow Type | $^{137}$Cs, Bq/kg of Air-Dried Mass | |
| --- | --- | --- | --- |
| | | Sod | Herbage |
| Loamy meadow-chernozemic | Moist floodplain | 3.0 | 0.6 |
| Loamy sand meadow | Normal dry | 10–14 | 2.0–3.0 |
| Loamy sand meadow | Moist floodplain | 12–15 | 8.0–11 |
| Loamy derno-podzolic | Normal dry | 4.0–14 | 1.0–4.0 |
| Sandy derno-podzolic | Normal dry | 40–63 | 5.0–9.0 |
| Sandy derno-podzolic | Dry, with excess moisture | 45–69 | 13–22 |
| Sandy derno-podzolic | Moist floodplain | 53–75 | 25–39 |
| Peat-gleic | Dry peat | 77–90 | 30–45 |
| Peat-gleic | Floodplain peat | 123–172 | 58–82 |
| Peat-gleic | Lowland peat | 170–198 | 135–189 |

**TABLE 5.6.4.**  Maximum $^{137}$Cs Soil Contamination Producing Internal
Dose of 1 mSv/yr

| Soil Type | Meadow (kBq/m$^2$) | Tilled Land (kBq/m$^2$) |
|-----------|--------------------|--------------------------|
| Peat-bog | 11–22 | 22–93 |
| Sandy podzolic | 74 | 130 |
| Light-gray podzolic | 148 | 185 |
| Loamy chernozem | 222 | 295 |

### 5.6.3 **Food Chain**

To understand the agricultural product contamination, we studied several food chains. This began with grasses grown on meadows and pasturelands as well as natural lands.

Contamination of natural ecosystems and cultivated meadow or pasture-land (seminatural) ecosystems has been monitored and mapped since 1987. A detailed survey of meadows and pastureland used for grazing of cattle on private farms was conducted at 200 collective farms and state farms through-out Ukraine. The results were used for certification of fodder-growing land; the most contaminated meadows and pastureland were removed from use, and specific fodder-management recommendations were provided for each farm. Meadows and pastureland in some regions were resurveyed in 1990–1992 to monitor compliance with the recommendations and determine their effectiveness.

During the acute phase of the Chornobyl accident, sorting of contaminated animals was generally not practiced; a large number of animals (more than 15,000 cattle in Ukraine alone) were slaughtered in the days immediately following the accident. Disposition of the carcasses created tremendous health and economic difficulties; naturally, holding the meat under refrigeration did not eliminate the long-lived radionuclides through physical decay. On the other hand, placing the animals on uncontaminated feed could remove radiocesium from the animals within 1–2 months [Prister et al. 1991].

The criticality of natural pastureland can clearly be demonstrated by comparing the $^{137}$Cs content of meat from domestic and wild animals. In the summer of 1993, the radionuclide content in meat from collective-farm cows which grazed on land that had been farmed was 20–44 Bq/kg. The radionuclide concentration in meat from private farm cows during this same time was 75–500 Bq/kg, while that for meat from wild animals was 500–1,540 Bq/kg [Prister et al. 1996b].

Berries and mushrooms collected from natural ecosystems may also make a significant contribution to the annual $^{137}$Cs intake for Polessye residents. Berries picked from uncultivated areas had contamination levels of 223 Bq/kg; mushrooms picked in uncultivated areas had levels up to 7,000 Bq/kg. For comparison, the mean radionuclide content in agriculturally grown plants in the population center of Dubrovitsya (Rivne Oblast) was 17–78 Bq/kg [Prister et al. 1996a].

### 5.6.4 Second Phase of Agricultural Product Monitoring

During the second phase of the accident, special attention was devoted to environmental and health monitoring aimed at identifying critical products in the human diet and the food chain for these products. Research in five oblasts (Volyns'ka, Rivne, Zhitomir, Kiev, and Chernigov) indicated that the critical food chains involved pastureland in forests and swamps. Monitoring allowed the identification of meadows and pastureland with high $^{137}$Cs contamination levels or high $^{137}$Cs hay-and-vegetation transfer coefficients [Prister et al. 1996b; Prister et al. 1996c].

Analysis of the $^{137}$Cs content of combined soil and hay samples from 20 personal farms in the village of Milyach (Rivne Oblast) showed that differences of as little as a factor of 2.2 in pastureland contamination could lead to differences of more than a factor of 100 in the $^{137}$Cs content of the grass. These differences continued throughout the entire grazing season.

As contamination of raw fodder on individual auxiliary farms is the main source of milk contamination under Polessye conditions, critical pastureland must be improved, cultivated, or removed from use.

### 5.6.5 Agricultural Countermeasures

Changing the land-use structure by relocating production of critical products to the least-contaminated and most productive land is a highly effective organizational countermeasure [Prister et al. 1993]. It should be implemented in combination with differential storage and use of feed characterized by various levels of radioactive contamination.

By 1992, Ukrainian agricultural countermeasures enabled the production of products with cesium content in excess of established standards to be virtually eliminated in the social sector. However, the radiation environment on individual auxiliary farms is much more complex because of the difficulty of eliminating natural and seminatural ecosystems from use for haying and grazing, especially in areas with peat-bog soils.

Care should be taken in the organization and implementation of counter-measures in the agricultural industry to distinguish between the following two goals: supplying uncontaminated products to the urban population and supplying uncontaminated products to the rural population which produces the products. While the countermeasures for the first category are largely aimed at reducing the collective dose, the countermeasures for rural inhabitants must still address the issue of preventing exposure in excess of the maximum permissible individual dose [Prister et al. 1996a].

5.6.5.1 **Crops and Countermeasures.** Agricultural countermeasures have been in place since the Chornobyl NPP accident, drawing on the experience gained during remediation of the 1957 accident in the southern Urals. Separate application of potassium and phosphorus fertilizer has been shown to be ineffective and may even lead to an increase in the transfer coefficient. Application of potassium fertilizers, especially in combination with other mineral nutrients with N:P:K ratios of 1:1.5:2 increases the harvest and significantly reduces (by a factor of 2–14) $^{137}$Cs uptake in the product [Prister et al. 1993].

Application of sand and clay in addition to chemical fertilizer reduces the transfer of contamination between peat soils and vegetation. The sand and clay increase the sorption resistance of the soil and reduces the soil-to-vegetation transfer coefficient by a factor of 2.5–5 [Prister et al. 1993].

When countermeasures are implemented over large areas under production conditions, countermeasure effectiveness is somewhat lower than under experimental conditions. The difference in effectiveness is largely because of failure to comply with procedures, and is approximately a factor of 1.5–3 for mineral soils and 3–16 on peat soils.

The use of chemical fertilizers to increase harvests and reduce product contamination is an economically beneficial countermeasure. Implementation of this countermeasure will reduce the individual doses received by residents who use the products thus produced. The total amount of radionuclides consumed, and therefore the collective dose, will be reduced by a smaller amount, and may even increase slightly. The collective dose can be managed in such cases by using the contaminated products as feed for livestock during the early stages of the fattening process.

5.6.5.2 **Livestock and Countermeasures.** Implementation of a three-stage fattening process for livestock, with successive reductions in the $^{137}$Cs content of the feed as the animal approaches commercial weight is a highly effective countermeasure. In this process, the final stage uses feed with as little contamination as possible for the final 4–10 weeks before the animals are slaughtered.

Various enteric adsorbents and feed additives based on clinoptilolite and other minerals have found successful application in livestock breeding. These materials have been used to reduce milk contamination by as much as a factor of 5 and meat contamination by as much as a factor of 3 under production conditions.

# 6

## здровъе риск

# INDIVIDUALS

## *Accident Remediation Personnel and Public Doses*

### I. Los' and V. Poyarkov

This chapter focuses on accident remediation staff (also known as "liquidators") and people living near the Chornobyl Nuclear Power Plant (NPP) at the time of and after the accident. The category "remediation personnel" includes firefighters who attempted to control the fires produced by the explosion and on the graphite fire as well as those who worked on other remediation tasks. The public includes people living in the 30-km Exclusion Zone near the reactor (who were evacuated shortly after the accident) and people living in contaminated areas, outside of the Exclusion Zone, that were defined by Ukrainian law in 1991. An enormous amount of scientific data has been obtained in the years since the accident, and new models have been developed for performing calculations, recovering retrospective doses, and predicting doses received by accident remediation personnel and the public.[1]

## 6.1 DOSES RECEIVED BY ACCIDENT REMEDIATION PERSONNEL

Nearly 300,000 people were involved in remediation of the Chornobyl NPP accident. For a variety of reasons, consistent, reliable dose information was not always collected on the workers. To determine the doses received by accident remediation personnel, we, the authors, have gone to two sources: the Ukrainian National Register and the Chornobyl Retrospective Dose Reconstruction Department.

---

[1]Editor's Note: For further discussion on the medical consequences of the Chornobyl accident, see Appendix B.

The Ukrainian National Register, which was established following the accident, lists approximately 175,000 remediation personnel [Buzunov et al. 1986], of whom 126,000 participated in remediation efforts during 1986 and early 1987, when accident remediation personnel were exposed to the highest radiation levels. The estimated mean exposure doses listed in the Register for personnel involved in accident remediation during 1986–1987 range from 120–180 mSv, and the existing (or retrospectively reconstructed) data follow a log-normal distribution (Figure 6.1.1). These data are not very reliable and are in need of further analysis and refinement.[2] One improvement method would involve examination, classification, and analysis of the existing personnel locator sheets and improvement of dose reconstruction methods [Ministry for Chornobyl Affairs 1996].

However, these data can be used to obtain a first-approximation estimate for the collective dose received by accident remediation personnel. The total collective dose for all personnel involved in remediation of the Chornobyl NPP accident in 1986–1987 was approximately 16,000 person-Gy.

Individual doses are given for only 22% of the 126,000 accident remediation personnel who worked in the 1986–1987 time period and are listed in the Ukrainian National Register. Given the log-normal distribution of the exposed population, one would expect that 6–15% of the approximately

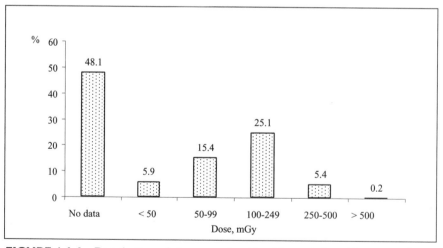

**FIGURE 6.1.1.** Distribution of doses received by Chornobyl accident remediation personnel in 1986–1987 [Buzunov et al. 1986]

---

[2]Editor's Note: The primary source of uncertainty in the individual and collective doses at this time is the lack of individual dosimetric monitoring. In many cases, dose control was achieved by establishing allowable stay times in high radiation areas and enforcing those stay times with either a timekeeper or personnel relay.

98,000 unmonitored accident remediation personnel in this group to have doses in excess of 250 mGy, and for half of these personnel (3–7.5%) in turn to have doses in excess of 500 mGy [Ministry for Chornobyl Affairs 1996].

The Chornobyl Retrospective Dose Reconstruction Department reconstructed the doses of plant personnel involved in emergency operations. Their estimated doses are based on detailed reconstruction of the routes taken by individual personnel relative to the physical environment and radiation environment, followed by various dose estimation calculations [Vasil'chenko et al. 1995].

Approximately 3,500 people worked at the plant during the accident remediation period (May–December 1986) and were subject to Chornobyl NPP dose accounting. According to Ukrainian National Register and Chornobyl's dose reconstruction group, the mean individual dose for these 3,500 people was 97 mGy.[3] During the accident (26–30 April), 1,600 people worked on remediation; their mean individual dose was 406 mGy. Eighteen of these individuals died of acute radiation sickness (with a mean individual dose 6,250 mGy) [Vasil'chenko et al. 1995]. Based on this information, the collective dose received by all of the accident remediation personnel for 1986 was 980 person-Gy (or including those who died, 1,086 person-Gy).

If we assume that all other accident remediation personnel (e.g., nonplant personnel) received similar doses and extrapolate this collective dose to all 126,000 personnel involved in accident remediation, this leads to a collective dose of approximately 40,000 person-Gy.[4]

## 6.2 DOSE RECEIVED BY THE PUBLIC

### 6.2.1 Collective Effective Doses

Protective countermeasures to be implemented are based on the magnitude and structure of the collective effective dose predicted to be received by the public. Thus, it is helpful to have a breakdown of the collective dose of accident-related radiation received by Ukrainians not working at the plant. The dose data are shown here for certain exposure variables: urban versus rural populations, for external and internal exposures, and by time since the accident.

Children who were under the age of 1 in 1986 are considered the "critical" population group, i.e., the group that will receive the largest lifetime

---

[3]Editor's Note: The normal annual occupational dose limit according to the State Sanitary Code of the USSR was 50 mSv (5 rem) for practical purposes 1 Gy = 1 Sv.

[4]Editor's Note: This assumption is likely to lead to an overestimation of the collective dose. A more detailed analysis by Illichev et al. (1996) concludes that the collective dose to accident remediation personnel was most likely on the order of 9,800 person-Gy.

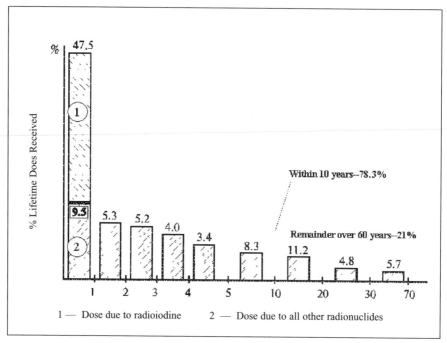

**FIGURE 6.2.1.** Estimated accumulation of accident-related effective doses for children born in 1986

accident-related dose. Figure 6.2.1 shows the cumulative accident-related dose as a function of time for the 70-year period from 1986–2056 for a child born in 1986. As the figure indicates, nearly 80% of the lifetime dose was accumulated during the first 10 years following the accident.

**Rural versus Urban Populations.** Figure 6.2.2 shows the dose structure for the rural population, averaged over all population centers within the zones where radiation monitoring was required by Ukrainian law. As the figure indicates, $^{137}$Cs is the main source of radiation affecting the public.

The dose received by the rural population makes up the bulk of the total collective dose (Figure 6.2.3) [Ministry for Chornobyl Affairs 1996]. Although the rural areas are less densely populated, the people there received significant exposures due to the consumption of food grown in private plots.

Figure 6.2.4 shows the breakdown of the total mean annual collective effective dose for the rural population over the 10-year period since the accident as a function of exposure type. Countermeasures are relatively ineffective against external exposure and most of this exposure occurred within a few years of the accident. Internal dose is controllable, and the collective

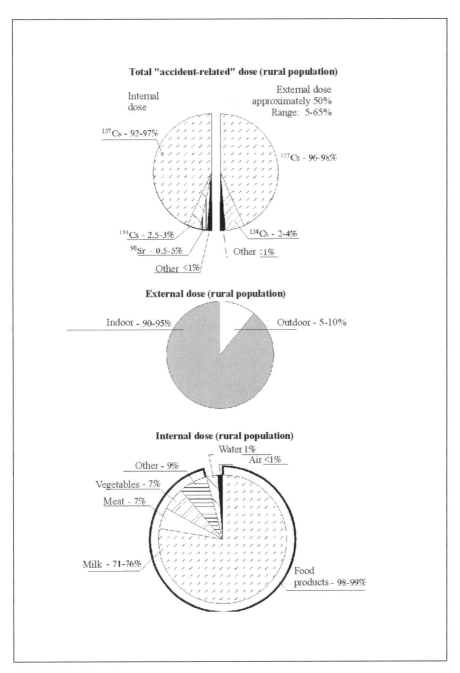

**FIGURE 6.2.2.** Structure of annual accident-related doses received by public in 1997 (estimated values averaged over radiation monitoring zones)

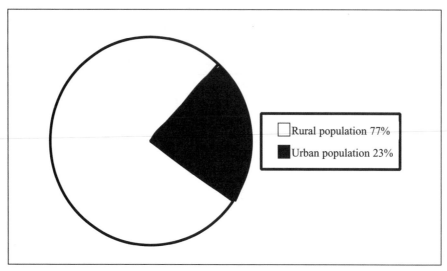

**FIGURE 6.2.3.** Ratio of collective total effective doses accumulated between 1986 and 1996 for people in urban and rural areas contaminated by the accident

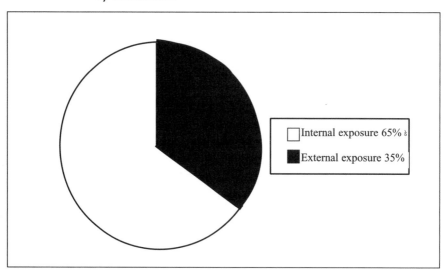

**FIGURE 6.2.4.** Ratio of total mean annual collective effective internal and external doses accumulated between 1986–1996 for rural areas contaminated by the accident

dose data should be used to optimize the effort toward reduction of internal exposure.

Recent estimates [Ministry for Chornobyl Affairs 1996] indicate that the total collective dose received by the 116,000 people who were evacuated from their homes and those who voluntarily left because of radioactive contamination from the accident (i.e., the exposed population) is approximately 47,500 person-Sv (not including thyroid exposure to radioiodine). Ninety-five percent of this dose is due to $^{137}$Cs, while 5% is due to $^{90}$Sr, and a fraction of a percent is due to plutonium.

**External versus Internal Dose.** External gamma radiation from accident-related radionuclides deposited on the surface was the main dose source during the first few years after the accident. Since the Chornobyl NPP accident, various subgroups of the public have been subjected to a total of more than 50,000 direct thermoluminescence-dosimetry measurements of external exposure; however, such measurements have only been performed on 1% of the exposed population. Because of this, dose reconstruction is performed using models of the spatial and temporal distribution of external gamma-ray doses.

Table 6.2.1 lists the estimated annual external doses received by Ukrainian residents over a 10-year period following the Chornobyl NPP accident as calculated using the model described by Likhtarev et al. [1996b]. The dose received by the public during May 1986 was equal to that received over the next 11 months. Half of the dose received over the first 6 years was received during the first year.

**TABLE 6.2.1.** Effective Dose of External Radiation Received by Ukrainians from Initial $^{137}$Cs Contamination, 1986–1996, $\mu$Sv/(kBq/m$^2$)

| Population Group | Year | | | | | | | | | | |
|---|---|---|---|---|---|---|---|---|---|---|---|
| | 1986* | 1987 | 1988 | 1989 | 1990 | 1991 | 1992 | 1993 | 1994 | 1995 | 1996 |
| Urban | 6.0 | 1.5 | 1.4 | 1.4 | 1.3 | 1.3 | 1.2 | 1.2 | 1.1 | 1.0 | 1.0 |
| Rural | 14.3 | 3.5 | 3.4 | 3.3 | 3.2 | 3.1 | 3.0 | 2.7 | 2.5 | 2.4 | 2.2 |

*The 1986 data was estimated after the accident.

**Time Since the Accident.** Figure 6.2.5 shows the total collective effective dose from the Chornobyl NPP accident as a function of time for the rural population in contaminated areas. The external accident-related dose was only of importance for the first 2 years following the accident. The initial value and change in this quantity with time were governed solely by the

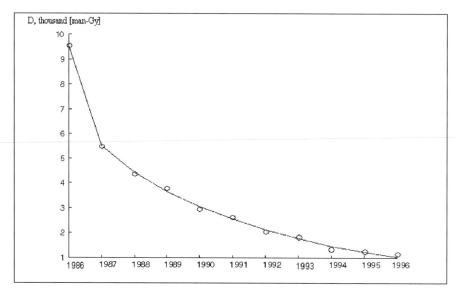

**FIGURE 6.2.5.** Total mean annual collective effective dose to rural Ukrainians from the accident as a function of time. The point for 1986 on the graph corresponds to the end of 1986.

initial density and decay of the $^{137}$Cs fallout. The collective internal dose shows quite different behavior and is largely determined by the soil-to-vegetation and soil-to-milk transfer coefficients, as well as the diet of the inhabitants in contaminated areas. This component of the accident-related dose has remained at a substantial level since the accident, which explains the additional attention it has received in countermeasure planning efforts.

### 6.2.2 Individual Effective Doses

The distribution of individual doses received by residents of areas contaminated as a result of the accident is highly nonuniform. Figure 6.2.6 shows the distributions of individual effective dose due to $^{137}$Cs for residents of two population centers in different oblasts determined via direct whole-body-counter measurements performed at the Ukrainian Academy of Medical Sciences Scientific Center for Radiation Medicine.

The dose burdens show a large amount of variability. Figures 6.2.7 and 6.2.8 show the mean annual internal and external effective doses from the accident to 2056 for residents of various population centers in the contaminated area. The graphs were constructed from currently available data, with predictions generated using existing mathematical models for prediction of the internal [Likhtarev et al. 1996b] and external [Likhtarev et al. 1996a] doses.

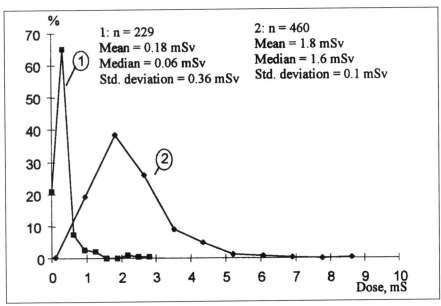

**FIGURE 6.2.6.** Frequency distribution of individual internal effective doses due to $^{137}Cs$ for people living in accident-contaminated areas (from direct measurements performed in 1995): 1—Village of Bazar, Narodichi Region, Zhitomir Oblast ($\sigma_{Cs} = 100$ kBq/m$^2$); 2—Village of Stare Selo, Rokitne Region, Rivne Oblast ($\sigma_{Cs} = 40$ kBq/m$^2$) [Ministry for Chornobyl Affairs 1996]

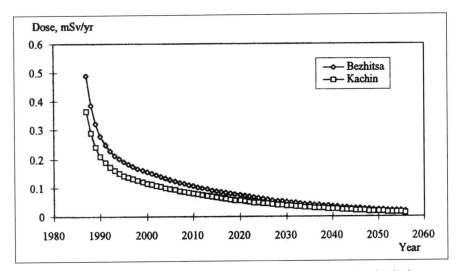

**FIGURE 6.2.7.** Behavior of received and predicted mean annual individual internal effective doses for two highly typical population centers in the contaminated area: village of Bezhitsa, Rokitne Region, Rivne Oblast ($\sigma_{Cs} = 100$ kBq/m$^2$); and village of Kachin, Kamin-Kashirskii Region, Volinska Oblast ($\sigma_{Cs} = 75$ kBq/m$^2$)

Table 6.2.2 lists the individual annual effective internal doses due to $^{137}$Cs (normalized to unit surface contamination) from 1986–1996 for all radiation-contaminated areas. These areas were classified based on standard values of the soil-to-milk transfer coefficient [Likhtarev et al. 1996a].

The density of $^{90}$Sr fallout on Ukraine is almost an order of magnitude lower than that of $^{137}$Cs outside the Exclusion Zone. Moreover, this ratio varies as a function of plume direction, but generally in such a way as to favor $^{137}$Cs, i.e., the amount of $^{90}$Sr per unit intake of $^{137}$Cs increases.

Sample studies show that most residents of Zhitomir Oblast have skeletal strontium burdens of 100–200 Bq, while the skeletal burden of $^{90}$Sr from global fallout is 50 Bq per adult skeleton [Ministry for Chornobyl Affairs 1996]. The contribution of $^{90}$Sr to the total internal dose increases monotonically with time. In 1994, the peak contribution of $^{90}$Sr to the total internal dose was 6% for adults and 16% for children who were at least 1 year old at the time of the accident [Ministry for Chornobyl Affairs 1996]. The dose received by the public from transuranic elements is non-zero and will be several dozen to several hundred times smaller than the dose received by the public from cesium and strontium over the 70-year period following the accident.

**TABLE 6.2.2.** Effective Internal Dose Received by Rural Ukrainians per Unit $^{137}$Cs Contamination for Areas with Different Soil-to-Milk Transfer Coefficients, $\mu$Sv/(kBq/m$^2$).

| Soil-to-Milk Transfer Coefficient, (Bq/L)/ (kBq/m$^2$) | Year | | | | | | | | | | |
|---|---|---|---|---|---|---|---|---|---|---|---|
| | 1986 | 1987 | 1988 | 1989 | 1990 | 1991 | 1992 | 1993 | 1994 | 1995 | 1996 |
| <1 | 9 | 7 | 5 | 5 | 3 | 3 | 2 | 2 | 1 | 1 | 1 |
| 1–5 | 42 | 34 | 27 | 21 | 17 | 14 | 10 | 9 | 6 | 6 | 5 |
| 5–10 | 95 | 75 | 60 | 48 | 37 | 30 | 24 | 19 | 15 | 12 | 10 |
| >10 | 176 | 139 | 110 | 88 | 70 | 55 | 44 | 35 | 28 | 22 | 18 |

### 6.2.3 Doses Received by Individuals Evacuated During 1986

One of the most burdensome impacts of the Chornobyl NPP accident was the need to evacuate members of the public from the 30-km Exclusion Zone. The evacuation began on 27 April with the city of Pripyat and quickly expanded to a zone 30 km around the nuclear plant.

The external gamma dose received by 90,000 people living in the 30-km Exclusion Zone was retrospectively estimated through analysis of approximately 40,000 questionnaires. The questionnaires allowed individual routes to be synchronized with the variations in the gamma-ray field at various population centers in the zone as a function of time and also enabled the amount of time spent outdoors to be taken into account [Likhtarev et al. 1994a; Chumak et al. 1989].

The range of doses received turned out to be quite large, from 1–660 mSv, although only 0.1% of the zone residents, or approximately 100 individuals, received doses greater than 250 mSv. However, this range does not take into account any additional dose received outside of population centers. The distribution of the population over the various dose subgroups is given in Table 6.2.3.

Analysis of the dose-frequency distribution and the nonuniformity of radioactive contamination in the zone indicates that the likelihood of a zone resident receiving a dose sufficient to trigger acute radiation sickness was $6.6 \times 10^{-5}$. The theoretical likelihood of acute radiation sickness for residents of Pripyat was virtually zero within the city, but migration outside the city limits would have increased the likelihood to comparable values, due to the closeness of the NPP and the very high radioactive contamination in areas adjacent to the NPP.

A breakdown of the mean individual external doses received by members of the public in Pripyat (population 49,360) and other parts of the 30-km Exclusion Zone (population 40,239) by age group is provided in Table 6.2.4.

The inhalation dose was estimated using a model to describe the transport of radioactivity in the form of fuel particles. The results of the dose calculations for the most heavily exposed organs (normalized to an exposure rate of 0.01 mGy/h) are provided in Table 6.2.5.

Because radioactive material was released in the form of fuel particles, residents of the Exclusion Zone could have inhaled several thousand of these

**TABLE 6.2.3.** Dose Subgroups as a Percentage of the Population of Pripyat and Other Population Centers in the 30-km Exclusion Zone

| Dose Subgroup, mGy | Pripyat, % (population 49,360) | 30-km Zone, % (population 40,239) |
|---|---|---|
| 0 - 50 | 98.58 | 86.17 |
| > 50 | 1.28 | 10.50 |
| > 100 | 0.14 | 3.20 |
| > 250 | — | 0.10 |
| > 500 | — | 0.03 |

**TABLE 6.2.4.** Estimated Mean Individual External Doses Received by Residents of Pripyat and Other Parts of the 30-km Exclusion Zone by Age Group, mSv [Chumak and Likhtarev 1996]

| Age Group, yr | Mean Individual External Dose | |
|---|---|---|
| | Pripyat | 30-km Zone |
| 0–3 | 8.2 | 7.7 |
| 3–7 | 9.0 | 8.5 |
| 7–12 | 9.4 | 10.2 |
| 12–16 | 11.2 | 13.4 |
| 16–25 | 13.0 | 16.7 |

**TABLE 6.2.5.** Estimates of Effective and Equivalent Individual Doses to the Lungs and Lower Large Intestine

| Time Before Evacuation, d | Effective Dose, Sv | | Equivalent Dose to Lungs, Sv | | Effective Dose to LLI, Sv |
|---|---|---|---|---|---|
| | $\beta$-emitters | $\alpha$- and $\beta$-emitters | $\beta$-emitters | $\alpha$- and $\beta$-emitters | $\alpha$- and $\beta$-emmiters |
| 1 | $1.1 \times 10^{-4}$ | $1.4 \times 10^{-4}$ | $6.9 \times 10^{-4}$ | $7.9 \times 10^{-4}$ | $4.3 \times 10^{-5}$ |
| 2 | $1.4 \times 10^{-4}$ | $1.7 \times 10^{-4}$ | $8.7 \times 10^{-4}$ | $1.0 \times 10^{-3}$ | $5.3 \times 10^{-5}$ |
| 3 | $1.6 \times 10^{-4}$ | $2.1 \times 10^{-4}$ | $1.0 \times 10^{-3}$ | $1.2 \times 10^{-}$ | $6.3 \times 10^{-5}$ |
| 4 | $1.9 \times 10^{-4}$ | $2.3 \times 10^{-4}$ | $1.2 \times 10^{-3}$ | $1.4 \times 10^{-3}$ | $7.1 \times 10^{-5}$ |
| 5 | $2.1 \times 10^{-4}$ | $2.6 \times 10^{-4}$ | $1.3 \times 10^{-3}$ | $1.5 \times 10^{-3}$ | $7.8 \times 10^{-5}$ |
| 6 | $2.2 \times 10^{-4}$ | $2.8 \times 10^{-4}$ | $1.5 \times 10^{-3}$ | $1.7 \times 10^{-3}$ | $8.3 \times 10^{-5}$ |
| 7 | $2.4 \times 10^{-4}$ | $3.0 \times 10^{-4}$ | $1.6 \times 10^{-3}$ | $1.8 \times 10^{-3}$ | $8.8 \times 10^{-5}$ |
| 8 | $2.5 \times 10^{-4}$ | $3.1 \times 10^{-4}$ | $1.6 \times 10^{-3}$ | $1.9 \times 10^{-3}$ | $9.1 \times 10^{-5}$ |
| 9 | $2.6 \times 10^{-4}$ | $3.3 \times 10^{-4}$ | $1.7 \times 10^{-3}$ | $2.0 \times 10^{-3}$ | $9.5 \times 10^{-5}$ |
| 10 | $2.7 \times 10^{-4}$ | $3.4 \times 10^{-4}$ | $1.8 \times 10^{-3}$ | $2.1 \times 10^{-3}$ | $9.8 \times 10^{-5}$ |
| 15 | $3.1 \times 10^{-4}$ | $4.0 \times 10^{-4}$ | $2.1 \times 10^{-3}$ | $2.5 \times 10^{-3}$ | $1.1 \times 10^{-4}$ |
| 20 | $3.6 \times 10^{-4}$ | $4.6 \times 10^{-4}$ | $2.5 \times 10^{-3}$ | $2.9 \times 10^{-3}$ | $1.3 \times 10^{-4}$ |

LLI = Lower large intestine

particles with an inhalable-fraction activity of up to 100 Bq. The Exclusion Zone residents were also exposed to a noninhalable-fraction activity of a few thousand becquerels.

The estimated risk of "hot-particle" inhalation for Pripyat residents, obtained using the model proposed by Repin et al. [1993], was approximately $2 \times 10^{-6}$. The risk for inhalation of particles of any type into the lungs was $1 \times 10^{-5}$.

Comparison of these risk values against the risk values obtained using the standard International Commission on Radiological Protection (ICRP) dosimetric model indicates that the ICRP model is fairly conservative, as it suggests that Pripyat residents have an exposure risk of $2 \times 10^{-3}$. These results provide a basis for ignoring the "hot particles" as an independent factor in determining the internal exposure.

The retrospectively estimated beta exposure for the skin and cornea was also obtained from the gamma-ray exposure rate and was determined as of the day of evacuation [Repin et al. 1993]. The distribution of beta exposures was determined for all population centers in the zone. It was shown that the maximum skin exposures might reach 1.0–1.7 Gy; i.e., the beta exposure could be three to four times the external gamma-ray exposure. The exposures to the cornea are generally less than 0.1 Gy; however, peak values run as high as 0.4–0.5 Gy.

Figure 6.2.9 summarizes the results on a common exposure scale. The rectangles in Figure 6.2.9 indicate the maximum or minimum exposure values recorded or calculated for each type of exposure, with the median values being indicated in black. In comparing the various types of exposures and exposure levels for individual organs, the thyroid gland (numbers 3–9 in Figure 6.2.8) stands out the most above the whole-body gamma-ray exposure background (2). The next most heavily exposed organ is the skin (12), with the lungs (10) third, and the exposures of the lower large intestine (11) and cornea (13) last. Figure 6.2.8 also shows the range of total exposures received outdoors in population centers before the evacuation (1).

The collective external gamma-ray dose for the people evacuated from the 30-km zone –approximately 1,300 person-Sv—was also calculated from the data on the mean doses received by the evacuated population in various age subgroups and the number of members of the public in each subgroup [Chumak and Likhtarev 1996].

### 6.2.4 Doses Received by People Resettled Since 1987

The beginning of the late resettlement period is marked by proclamation of the Ukrainian Law on the Status and Social Protection of Citizen Victims of the Chornobyl Disaster in 1991. This law defined several radioactive contamination zones (see Table 6.2.5). Since adoption of the law in 1990, approximately 20,000 people have been resettled from the mandatory resettlement zone and 32,000 people have been resettled from the voluntary resettlement zone under this law.

Because of the lack of information on the actual resettlement figures for each individual year, it was assumed that an identical number of residents were resettled from 1991–1997. Information on the specific population cen-

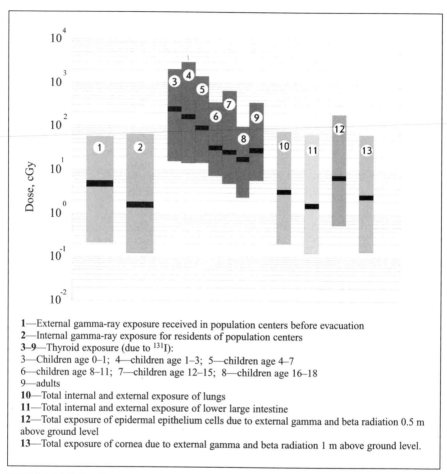

1—External gamma-ray exposure received in population centers before evacuation
2—Internal gamma-ray exposure for residents of population centers
3–9—Thyroid exposure (due to $^{131}$I):
3—Children age 0–1;  4—children age 1–3;  5—children age 4–7
6—children age 8–11;  7—children age 12–15;  8—children age 16–18
9—adults
10—Total internal and external exposure of lungs
11—Total internal and external exposure of lower large intestine
12—Total exposure of epidermal epithelium cells due to external gamma and beta radiation 0.5 m above ground level
13—Total exposure of cornea due to external gamma and beta radiation 1 m above ground level.

**FIGURE 6.2.9.** Composite median and range of whole-body and individual-organ dose for 30-km Exclusion Zone residents after the accident (up until time of evacuation)

ters resettled in any given year does not exist; so, mean contamination values were used for the mandatory and voluntary zones. Using the effective equivalent internal and external exposures per unit radiocesium fallout from *Ukrainian National Report on the 10th Anniversary of the Accident at ChNPP* [Ministry for Chornobyl Affairs 1996], the estimated accident-related collective effective dose received by residents of the mandatory resettlement zone before resettlement was 481 person-Sv, while that received by residents resettled from the voluntary resettlement zone was 1,280 person-Sv. The collective accident-related dose for all resettled residents was on the order of 1,760 person-Sv.

**TABLE 6.2.5.** Definition Criteria for Boundaries of Radioactive Contamination Zones

| Zone Category and Description | Defining Criteria for Boundary of Zone |
|---|---|
| 1. Exclusion Zone | Area evacuated in 1986 (unchanged) |
| 2. Mandatory resettlement zone | $\sigma_{Cs} > 555$ kBq/m$^2$ or $\sigma_{Sr}$  111 kBq/m$^2$ or $\sigma_{Pu}$  3.7 kBq/m$^2$, with $D^*_{eff} > 5$ mSv/yr |
| 3. Voluntary resettlement zone | $185 \leq \sigma_{Cs} \leq 555$ kBq/m$^2$, $5.5 \leq \sigma_{Sr} \leq 111$ kBq/m$^2$, $0.37 \leq \sigma_{pu} \leq 3.7$ kBq/m$^2$, with $D^*_{eff} > 1$ mSv/yr |
| 4. Zone of increased radioenvironmental monitoring | $37 \leq \sigma_{Cs} \leq 185$ kBq/m$^2$, $0.74 \leq \sigma_{Sr} \leq 5.5$ kBq/m$^2$, $0.185 \leq \sigma_{Pu} \leq 0.37$ kBq/m$^2$, with $D^*_{eff} > 0.5$ mSv/yr |

$D^*_{eff}$ is the effective *calculated* dose in *addition* to that received before the accident.

## 6.2.5 Iodine Exposure

Exposure of the thyroid gland to radioactive iodine isotopes in the public, and children in particular, was one of the most serious public health impacts of the Chornobyl NPP accident. The contribution from the short-lived iodine isotopes (e.g. $^{133}$I, $^{134}$I, $^{135}$I, and $^{132}$Te) is significant for people who were living near the NPP and were evacuated early (e.g., the city of Pripyat and nearby villages). This category of individuals was typically exposed to radioiodine through inhalation. The fraction of dose arising from inhalation of short-lived radioiodine is estimated to be an additional 4–30% of the dose from inhalation of $^{131}$I.

Approximately 150,000 direct measurements of the $^{131}$I concentration in the thyroid gland were performed in May–June 1986 [Likhtarev et al. 1995]. These measurements could be divided into two quality categories: "undefined quality" and "high quality" (approximately 40,000 of the measurements) [Likhtarev et al. 1995]. These data were used to determine the mean thyroid exposure in the regions of Ukraine with the highest levels of radioactive contamination. See Figure 6.2.10 for detailed information on the exposure to residents of cities and villages.

The contribution of $^{131,133,135}$I and $^{132}$Te to the thyroid exposure was estimated relative to the $^{131}$I dose [Likhtarev et al. 1992]. The maximum additional thyroid exposure due to direct inhalation of $^{132}$I, inhalation of $^{132}$Te, and inhalation of ambient $^{132}$I formed from $^{132}$Te during the first 4 days was

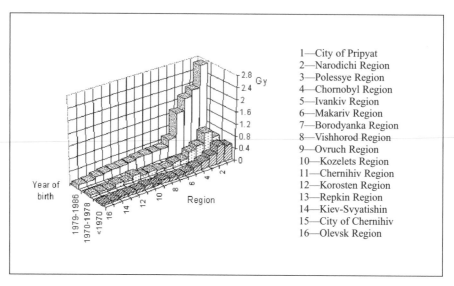

**FIGURE 6.2.10.** Measured data for mean thyroid doses

no more than 1.3% of the [131]I exposure. The contribution of [135]I to the total thyroid exposure (for a maximum exposure time of 5 hours) may be as much as 3–5% of that due to [131]I. On the other hand, [133]I may be the source of as much as 30% of the dose when the period of time the victim is in contact with the radioactive iodine isotopes is limited.

The main exposure pathway for members of the public who lived in distant areas and were not resettled during the iodine phase of the accident was through dairy products. In this case, the contribution of inhalation to thyroid exposure was less than 10%, and the fraction of this due to short-lived iodine isotopes was even smaller. For this population, the thyroid exposure was almost completely due to [131]I.

Analysis of the results of 10,600 direct measurements of [131]I concentrations in the thyroids of children and adults from the 30-km Exclusion Zone enabled the range of thyroid exposures to be determined. Further, the factors determining thyroid exposure were identified [Repin 1996; Goulko et al. 1996]. The exposure ranges are a strong function of age: 40–60% of children up to 3 years of age in the city of Pripyat had exposures of 2 Gy or higher, only 20% of children aged 4–7 had exposures of 2 Gy or higher, and less than 4% of adults had exposures greater than 2 Gy (see Table 6.2.6).

Iodine preparations taken by some residents as iodine prophylaxis according to the questionnaire responses received by Pripyat residents [Likhtarev et al. 1996a] provided some protection. For example, two doses

**TABLE 6.2.6** Thyroid Doses for 30-km Exclusion Zone Residents in Various Age Groups Based on Direct Measurements

| Age Group, yr | Median Value, Gy | Number of Measurements | Percentage with Exposures Greater than 2 Gy |
|---|---|---|---|
| < 1 | 2.48 | 100 | 57 |
| 1–3 | 1.68 | 1,228 | 42.6 |
| 4–7 | 0.8 | 1,785 | 19.9 |
| 8–11 | 0.26 | 2,679 | 5.41 |
| 12–15 | 0.17 | 3,107 | 2.37 |
| 16–18 | 0.15 | 726 | 2.07 |
| > 18 | 0.25 | 1,065 | 3.62 |
| Total number of measurements | | 10,676 | |

of stable iodine enabled Pripyat residents to reduce the expected exposure by a factor of 2. However, only 73% of the 854 Pripyat residents questioned took the iodine preparations [Repin 1996].

Direct measurements of thyroid activity in May–June 1986 covered 30–90% of the children and 1–10% of the adults living in several population centers. However, some population centers were not included in the measurements. A special thyroid dose-reconstruction project for residents affected by the accidental release was therefore begun (and is still under way) [Likhtarev et al. 1996a; Likhtarev et al. 1994b].

Retrospective dose reconstruction using a conservative "single-iodine-intake" model was performed. The study indicates that 79,500 of the 89,000 children (90%) in the eight most heavily contaminated regions and the city of Pripyat had thyroid exposures of 2 Gy or less (the threshold for deterministic effects on the thyroid) and 38,000 (40%) had low exposures (no greater than 0.3 Gy) [Likhtarev et al. 1993]. The eight regions were Chornobyl, Polessye, and Ivankiv Regions in Kiev Oblast; Kozelets, Repkin, and Chernihiv Regions in Chernihiv Oblast; and Narodichi and Ovruch Regions in Zhitomir Oblast. Using a more realistic model that takes into account the individual duration of iodine intake (data obtained through a questionnaire [Likhtarev et al. 1996c]), the exposure distribution is shifted to lower values (56,200 of the children and 328,000 out of 408,000 adults had exposures less than or equal to 0.3 Gy) [Likhtarev et al. 1993]. The more conservative model suggests that the collective dose to the thyroid received by this contingent of children was 116,000 person-Gy, while the realistic model yields 64,000 person-Gy. The corresponding values for the adults are 369,000 and 127,000 person-Gy [Likhtarev et al. 1993].

Estimates of the mean individual and collective doses were obtained for residents of Kiev in five age groups based on year of birth as follows: 1983–1986, 1979–1982, 1975–1978, 1971–1974, and before 1971 [Sobolev et al. 1996]. The corresponding individual doses by age group were as follows: 104, 62, 19, 18, and 41 mGy. The collective dose for Kiev residents was 83,000 person-Gy for the adult population, and 38,000 person-Gy for the nonadult population (under the age of 18). The collective dose to the thyroid for the entire nonadult population of Ukraine (i.e., the population less than 18 years of age at the time of the accident) was 400,000 person-Gy [Sobolev et al. 1996].

Individualization of the estimated group exposures for the thyroid gland is a special problem, which may be solved by analyzing various behavioral factors as a function of time. Such research is currently being performed using a specialized questionnaire (to which approximately 23,000 responses have been received) [Likhtarev et al. 1996a].

## 6.3 DOSES DUE TO NONACCIDENT-RELATED RADIATION SOURCES

To effectively study the radiation doses from the accident, we examined nonaccident sources of radiation, i.e., fallout from past atmospheric nuclear tests and natural sources such as cosmogenic and terrestrial radionuclides and radon.

### 6.3.1 Exposure to Global Fallout from Nuclear Weapons Testing

Radioactive fallout from atmospheric nuclear tests has been an important source of radiation exposure to the general public. Testing was conducted on the largest scale, and the release of radioactive products into the atmosphere was highest, in 1954–1958 and 1961–1962. The total yield of the tests conducted as of 1981 was 545.5 MT (trinitrotoluene equivalent).

Following an air burst, approximately 50% of the resulting radioactive products fall back to land or water in the vicinity (i.e., within 100 km) of the test. The remainder are released into the troposphere and stratosphere.

Personal exposure to radioactive products produced in nuclear weapons tests includes both internal exposure (inhalation of radionuclides with ground-level air and ingestion of radionuclides with food and water) and external exposure (radiation from radionuclides in ground-level air and on the surface of the ground). The mean effective individual annual dose is approximately 6 $\mu$Sv (as of 1996) [UNSCEAR 1993].

### 6.3.2 **Exposure to Sources of Natural Origin**

Radiation exposure from natural sources can be divided into three categories: cosmic rays and radiation of cosmogenic origin, radiation from the Earth, and radon. The mean total individual effective dose received by Ukrainian public from natural sources is 4.88 mSv/yr [Ministry for Chornobyl Affairs 1996].

### 6.3.2.1 **Cosmic Rays and Radiation from Radionuclides of Cosmogenic Origin.** Given that the average Ukrainian has a behavior factor (the age- and occupation-weighted relative time spent by an individual indoors and outdoors during the course of the day) of 0.8, the annual effective dose directly received from ionizing cosmic rays is estimated to average 240 $\mu$Sv/yr (under the assumption that the dose value is independent of altitude). The dose from the neutron component is approximately 30 $\mu$Sv/yr. The corresponding world-averaged effective doses are 300 and 80 $\mu$Sv/yr, respectively [UNSCEAR 1993].

The most important radionuclide formed as a result of interaction between atmospheric atoms and cosmic rays (cosmogenic atoms) is $^{14}$C. The mean annual individual effective dose to the world's population (including Ukraine) due to $^{14}$C is 12 $\mu$Sv/yr [UNSCEAR 1993] (based on a value of 230 Bq/kg of $^{14}$C in coal, and, therefore, annual emissions equal to $1 \times 10^{12}$ Bq of $^{14}$C). The contributions from the other cosmogenic radionuclides can be ignored.

Thus, the mean individual annual effective dose to Ukrainian public from cosmic rays is approximately 282 $\mu$Sv/yr. If we take into account the fact that Ukraine has a mean altitude slightly above sea level and use a screening factor of 0.82, the mean annual dose will be 288 $\mu$Sv/yr [Krisyuk 1989]. The value quoted by [Ministry for Chornobyl Affairs 1996] is 300 $\mu$Sv/yr. In any case, these estimates are only 6% apart, while the difference relative to the mean worldwide value (380 $\mu$Sv/yr [UNSCEAR 1993]) is 28%.

### 6.3.2.2 **Natural Radionuclides of Terrestrial Origin.** People are also exposed to gamma rays from radionuclides of natural origin both indoors and outdoors, such as radionuclides found in rocks, soil, and plants.

The outdoor mean annual individual effective dose has been calculated by a variety of sources; estimates range from 150 $\mu$Sv/yr [Krisyuk 1989; Ministry for Chornobyl Affairs 1996] to 280 $\mu$Sv/yr [Marei et al. 1984]. Thus, there are significant differences in the estimated mean individual outdoor dose to Ukrainian public from natural radionuclides of terrestrial origin. We believe that the results in *Ukrainian National Report on the 10th Anniversary of the Accident at ChNPP* [Ministry for Chornobyl Affairs 1996] (150 $\mu$Sv/yr) are more reliable, as they were averaged over Ukraine alone, while the results in

*Indoor Radiation Background* [Krisyuk 1989] and *Communal Radiation Hygiene* [Marei et al. 1984] are for the entire former Soviet Union.

The indoor external gamma-ray dose estimates range from 0.41 mSv/yr [Krisyuk 1989] to 0.46 mSv/yr [UNSCEAR 1993]. This suggests that the additional external dose from construction materials is approximately 0.14 mSv/yr, although *Ukrainian National Report on the 10th Anniversary of the Accident at ChNPP* [Ministry for Chornobyl Affairs 1996] quotes a value of 0.26 mSv/yr. For the same reasons mentioned above, the 0.26 mSv/yr value [Ministry for Chornobyl Affairs 1996] appears to be the better value for the mean individual indoor dose from natural radionuclides of terrestrial origin.

The mean individual effective internal dose due to natural radionuclides of terrestrial origin is 0.23 mSv/yr [Sobolev et al. 1996], of which 0.17 mSv/yr is due to $^{40}$K and 0.06 mSv/yr is due to uranium- and thorium-series radionuclides. Other sources quote 0.20 mSv/yr for the annual individual internal dose due to natural beta emitters [Ministry for Chornobyl Affairs 1996]. The available data cover a range of approximately 15%, i.e., they are fairly consistent.

6.3.2.3 **Radon.** When radium in rocks and soil decays, radon is produced. This tasteless, odorless gas can seep into buildings through basement cracks, sewer openings and joints between walls and floors. Radon breaks down into other radioactive elements that cling easily to airborne particles, such as dust and smoke.

The mean weighted individual effective dose to the Ukrainian public from $^{222}$Rn is 3.8 mSv/yr (1996 data) [Ministry for Chornobyl Affairs 1996]. The individual dose is highest in Ternopil Oblast (6.7 mSv/yr) and lowest in Volins'ka Oblast (1.34 mSv/yr). Los' [1993] quotes a value of 3.4 mSv/yr for the mean weighted individual annual dose from radon in Ukraine (earlier estimate from 1993). The UN Scientific Committee on the Effects of Atomic Radiation quotes a value of 1.2 mSv/yr for the world mean individual effective dose from radon [UNSCEAR 1993]. The value of 3.8 mSv/yr is consistent with the mean doses from radon in Western European countries. Ukraine is among the countries with a high radon dose rate. The only European countries with higher radon doses are Finland, Sweden, France, and Spain [Green et al. 1991].

### 6.3.3 Collective Dose Received from Nonaccident-Related Sources and Comparison with Accident-Related Sources

The collective doses from nonaccident-related radiation sources (both annual doses and doses for the 1986–1996 time period) were calculated for individuals evacuated and resettled in 1986–1987, individuals resettled after 1990, accident remediation personnel, and people continuing to live in areas contaminated with accident-related radionuclides. The calculation made use of data on the sources and magnitudes of the individual doses received by Ukrainian public from $^{222}$Rn in interior air [Pavlenko 1996], from water-borne $^{234,238}$U, $^{226}$Ra, and $^{222}$Rn in various regions of Ukraine [Ministry for Chornobyl Affairs 1996; Pavlenko et al. 1996], from medical exposure, and the Ukrainian-averaged mean annual dose due to radiation sources of natural origin (4.88 mSv/yr) [Ministry for Chornobyl Affairs 1996]. The results of these calculations are shown in Table 6.3.1.

**TABLE 6.3.1.** Total Annual Collective Doses Received due to Nonaccident-Related Sources[a]

| Group of Accident Victims | Annual Total Dose, person-Sv/yr | Number of Individuals in Group |
|---|---|---|
| Individuals evacuated in 1986 | 487 | 89 |
| Accident remediation personnel at Chornobyl NPP 1986 - 1987 | 684 | 126 |
| Individuals resettled from radioactive contamination Zones II and III after 1990[b] | 283 | 52 |
| Zone II | 109 | 20 |
| Zone III | 174 | 32 |
| Residents of contaminated zones (Zones II, III and IV) [a] | 10,4558 | 2,175 |
| Zone II | 116 | 24 |
| Zone III | 2574 | 549 |
| Zone IV | 7,869 | 1,602 |
| Total | 12,011 | 2,442 |

(a) As of 1995, the collective dose values were determined by the number of individuals in each group and therefore vary.

(b) For definitions of the zones, see Table 6.2.5.

The collective accident-related doses (except for radioiodine) for various groups of accident victims were compared against the doses received from sources of nonaccident-related origin from the time of the accident in 1986–1996. The total nonaccident-related collective dose received by the population of Ukraine (approximately 50 million persons) over this 10-year period was approximately 2,715,000 person-Sv, while the total accident-related dose received by the population of Ukraine was estimated to be 47,500 person-Sv [Ministry for Chornobyl Affairs 1996]. Thus, the former dose is more than 57 times the latter. An example of the difference in contamination by origin can be seen in the Zhitomir Oblast (population 1,495,900). In this case, the nonaccident-related dose is more than seven times the accident-related dose. In the future, this ratio will turn even more in favor of the nonaccident component of the post-accident total and annual dose received by all groups of victims. For example, the total collective dose received by the population of areas contaminated by the Chornobyl NPP accident from nonaccident-related sources during 1996 (approximately 10,550 person-Sv) was more than 13 times the collective "Chornobyl" dose for that same time period (approximately 760,000 person-Sv). During the time accident remediation personnel worked in the Exclusion Zone in 1986–1987, they received a collective dose (various estimates of which range from 16,000–40,000 person-Gy) that was several dozen times the annual collective dose from nonaccident-related sources for that population contingent.

## 6.4 RISKS FROM ACCIDENT-RELATED EXPOSURE AND NONACCIDENT-RELATED SOURCES

In evaluating the radiation risk, we considered individuals evacuated and resettled in 1986–1987, individuals resettled after 1990, accident remediation personnel, and people continuing to live in areas contaminated with accident-related radionuclides. Table 6.4.1 lists the accident-related collective doses (over a 10-year period) for these groups.

The radiation risks were calculated using the recommendations provided in ICRP Publication 60 [ICRP 1991]. A linear/quadratic dose-effect relation was used, with a mean age-weighted coefficient of 0.1 $Sv^{-1}$ for the risk of developing a fatal cancer due to radiation exposure in the dose range where somatic radiation effects appear. A factor of 2 was used for the extrapolation to small doses (less than 0.2 Sv, according to the United Nations Special Commission on the Effects of Atomic Radiation criterion), e.g., the risk of fatal cancer in this region is 0.05 $Sv^{-1}$ (or 0.04 $Sv^{-1}$ for adult workers).

The risk of developing a thyroid cancer from radioiodine was estimated using the calculations performed by Zvonova et al. [1991], which were based on the reduced-risk models used by the US National Council on Radiation Protection and Measurements [NCRP 1985]. For adults and children in contaminated regions, this risk is 8 and 46 cases of fatal thyroid cancer per 10,000 person-Gy, respectively.

Table 6.4.2 presents the results of the risk assessment for the development of fatal cancer (or thyroid cancer for those exposed to radioiodine). Fatal cancers were selected on the basis of the most precise diagnosable actual cause of death.

**TABLE 6.4.1.** Accident-Related Collective Doses Over a 10-Year Period for Exposed Population

| Exposed Population | Collective Dose, person-Sv | Number in Contingent, thousands |
|---|---|---|
| Population of Ukraine | 47,500 | 50,000 |
| Individuals evacuated in 1986 | 1,300 | 89 |
| Individuals resettled from Zones II and III after 1991 | 1,760 | 52 |
| Accident remediation personnel during 1986 - 1987 | 40,000 | 126 |
| Individuals whose thyroids were exposed to radioiodine: | | |
| The eight regions with the heaviest contamination[a] | 191,000[b] | 497 |
| Children in Kiev (age 0–18 in 1986) | 28,000[b] | 686 |
| Adults in Kiev | 83,000[b] | 1,968 |
| Total nonadult population of Ukraine as of 1986 | 400,000[b] | 13,183 |

(a) Chornobyl, Polessye, and Ivankiv Regions in Kiev Oblast; Kozelets, Repkin, and Chernihiv Regions in Chernihiv Oblast; and Narodichi and Ovruch Regions in Zhitomir Oblast.

(b) Collective dose in man-Gy

**TABLE 6.4.2.** Expected Number of Fatal Cancers as a Function of Accident- and Nonaccident-Related Exposure During the 10 Years Since the Accident

| Exposed Population | Expected Number of Cases per 100,000 Individuals | |
|---|---|---|
| | Accident-Related Exposure | Nonaccident-Related Exposure |
| Population of Ukraine | 5 | 272 |
| Individuals evacuated in 1986 | 73 | 272 |
| Individuals resettled from Zones II and III after 1991* | 169 | 272 |
| Accident remediation personnel during 1986 - 1987 | 1,270 | 272 |

*For definitions of Zones II and III, see Table 6.2.5.

In Section 6.3, we estimated the annual collective doses for various population groups due to nonaccident-related radiation sources. These collective doses were used to determine the risk for development of fatal cancers among selected population groups. For all of these groups of victims, the expected number of fatal cancers due to nonaccident-related radiation that will develop over the course of 10 years' exposure is 272 per 100,000.

The above results can be used to compare the risks of fatal cancers. As Table 6.4.2 clearly indicates, only for accident remediation personnel does the expected number of fatal cancers from accident-related radiation exceed that for radiation from nonaccident-related sources (by a factor of 4.7). For all other groups of victims, the number of fatal cancers expected is between 3 and 50 times lower than that for nonaccident-related exposure.

## 6.5 COMPARISON OF RADIATION RISK AGAINST OTHER NONRADIATION-RELATED SOURCES

In general, comparative risk assessment should not be used to determine acceptable levels of risk. There are several reasons for this. First, each hazard has different uncertainties, which vary over a wide range, depending on the hazard involved. Second, the quantitative exposure parameters (observations) leading to a given risk level are different for each category of hazards. Third, the fraction of the population susceptible to a given risk factor for each hazard will be different (generally speaking), and the comparison is based solely on the mean value. Fourth, and perhaps most important, simple risk comparisons do not take into account any qualitative characteristics of the hazards. For example, most people are afraid of the risk of being hit by lightning, even though the risk is very low. Some risks, such as the use of motor vehicles, are "forced risks." On the other hand, people sometimes consciously engage in high-risk activities, such as smoking.

The mean risk of death due to stochastic radiation effects for individuals who participated in the Chornobyl NPP accident remediation effort in 1986–1987 is estimated to be approximately $1.3 \times 10^{-2}$. This is equivalent to 1,600 cases of incurable radiation-induced cancer among the 126,000 individuals involved in accident remediation during 1986–1987.

However, the risk of accident remediation personnel developing incurable radiation-induced cancer and the appearance of severe hereditary effects within the next two generations is estimated to be $5.1 \times 10^{-3}$ (approximately 640 cases within the cohort of remediation personnel under consideration). This value was obtained using the nominal coefficients for the likelihood of stochastic radiation effects at a dose of 1 Sv proposed in ICRP Publication 60 [ICRP 1991] for adult workers: $0.8 \times 10^{-2}$ $Sv^{-1}$ for incurable cancer and $0.8 \times 10^{-2}$ $Sv^{-1}$ for hereditary effects.

These radiation effects should occur following some minimum latency period that will be different for various cancerous diseases and is estimated to average 5–10 years. Eighty-six leukemia cases have already been identified among accident remediation personnel from 1987–1993 [Buzunov et al. 1986].

# 7
## СОЦИСОЛГ РИСК

# SOCIETY
## *Social Risks After the Accident*

V. Poyarkov, V. Kholosha, and Yu. Saenko

In this chapter we offer some brief observations and analysis of the social risk posed by the Chornobyl Nuclear Power Plant (NPP) accident, particularly how it affected two groups of people: those who evacuated the resettlement zones and those who remained in or returned to the resettlement zones.

## 7.1 ORIGIN OF SOCIAL RISKS IN THE POST-CHORNOBYL ENVIRONMENT

Analysis of the social risks posed by Chornobyl is one of the more difficult areas of research because the data are often much more qualitative than quantitative. To understand the disruption caused by the accident, we must first set the social backdrop for Ukraine at the time of the accident. In the spring of 1986, Ukraine was still under Soviet rule, and authority for operation of the plant essentially resided with officials in Moscow. The physical distance of the authorities and their reluctance to share information related to the nuclear industry contributed to the lack of accurate information made available immediately following the accident.

The psychological impact of the Chornobyl accident resulted primarily from a lack of public information, particularly immediately after the accident, the stress and trauma of relocation, the breaking of social ties, and the fear that any radiation exposure is damaging and could damage people's health and their children's health in the future. It is understandable that people who were not told the truth for several years after the accident continue to be skeptical of official statements and to believe that illnesses of all kinds that now seem more prevalent must be due to radiation. The distress caused by this misconception of radiation risks is extremely harmful to people.

The lack of consensus about the accident's consequences and the way in which they have been politicized has led to psychological stress among the populations that are extensive, serious, and long lasting. Severe effects include feelings of helplessness and despair, leading to social withdrawal and loss of hope for the future. The effects are being prolonged and exacerbated by protracted debates over radiation risks, countermeasures and general social policy, and the occurrence of thyroid cancers attributed to the early exposures.

Between 1990 and the end of 1995, decisions were made by authorities to further resettle people in Ukraine (about 53,000 people), Belarus (about 107,000 people) and Russia (about 50,000 people). Evacuation and resettlement has created a series of serious social problems linked to the difficulties and hardships of adjusting to new living conditions.

Demographic indicators in contaminated regions have worsened: the birth rate has decreased, and the workforce is migrating from contaminated to uncontaminated areas, creating shortages of labor and professional staff.

The control measures imposed by the authorities to limit radiation exposure in contaminated territories have limited industrial and agricultural activities. Moreover, the attitude of the general population toward products from contaminated areas makes it difficult for produce to be sold or exported, leading to a reduction in local income.

Restrictions on people's customary activities make everyday life difficult and distressing. Major rehabilitative actions have been undertaken over the past years. However, it is necessary to provide the public with more and better information on the measures taken to limit the consequences of the accident, on present radiation levels, and on radionuclide concentrations measured in foodstuffs.

The social and economic conditions of people living and working in contaminated territories are heavily dependent on public subsidies. If the compensation system in force were to be reconsidered, some of these funds could be redirected to new industrial and agricultural projects.

Exacerbated by the political, economic, and social changes of the past years, the consequences of the Chornobyl NPP accident and the measures taken in response have led to a worsening in the quality of life and of public health and to unfavorable effects on social activity. This situation was further complicated in the years after the accident by incomplete and inaccurate public information on the accident's consequences and on measures for their alleviation.

## 7.2 SOCIAL SATISFACTION INDEX OF THE VICTIM POPULATION

In March 1997, approximately 1,200 respondents were surveyed to determine their level of satisfaction with basic life parameters (i.e., the social satisfaction index), providing an approach for numerical determination of subjective risk (Table 7.2.1). The survey focused on four groups of people:

- residents of Zone 2 (mandatory resettlement zone)[1] that have never been evacuated or resettled
- residents of Zone 3 (voluntary resettlement zone) that have never been evacuated or resettled
- resettled evacuees
- residents of a "clean" area used as a control population.

The level of satisfaction with basic life parameters (0-100 points) was assessed on the following scale:

- Completely unsatisfied: 0
- Moderately unsatisfied: 20
- Satisfied: 50
- Highly satisfied: 75
- Completely satisfied: 100

These data indicate that there is basically no difference between the Zone 2 and 3 residents in terms of level of satisfaction. Neither group has experienced the upheaval and stress of relocation and accompanying changes in work and social life. It is noteworthy that the residents of these areas generally express a higher satisfaction than either the resettled evacuees or the control group.

In conclusion, social risks, particularly the subjective component of these risks, is an important part of understanding the impact of the Chornobyl NPP accident. This problem is important not only for understanding the post-Chornobyl behavior of victims, but also for analyzing public behavior relative to phenomena such as smoking where there is clearly a relationship between involuntary risk, voluntary risk, objective risk, and subjective risk. This research has been only a first attempt at understanding how this new field of study applies to the Chornobyl accident. Clearly this area deserves further study.

---

[1]Editor's note: Zone 2 areas are those where the density of radionuclides in the soil exceeds the following levels: cesium isotopes of 15 Ci/km$^2$, strontium isotopes of 3.0 Ci/ km$^2$, and plutonium isotopes of 0.1 Ci/km$^2$. Zone 3 is defined by the following contamination levels: cesium isotopes of 5.0 - 15.0 Ci/ km$^2$, strontium isotopes of 0.15 - 3.0 Ci/ km$^2$, and plutonium isotopes of 0.01 - 0.1 Ci/ km$^2$.

**TABLE 7.2.1.**  Satisfaction with basic life parameters (Social Satisfaction Index) (estimated reliability ± 5 points)

| Categories and Parameters of Social Satisfaction | Zone 2 Residents | Zone 3 Residents | Resettled Evacuees | Control Group |
|---|---|---|---|---|
| **1. *Health*** * | 37 | 39 | 29 | 37 |
| 1.1. Physical health | 30 | 33 | 26 | 40 |
| 1.2. Psychological health | 59 | 61 | 38 | 44 |
| 1.3. Medical services | 21 | 24 | 23 | 26 |
| **2. *Family*** * | 82 | 83 | 52 | 65 |
| 2.1. Family well-being and harmony | 82 | 83 | 52 | 65 |
| **3. *Welfare*** * | 46 | 52 | 38 | 42 |
| 3.1. Income and financial situation | 19 | 18 | 22 | 25 |
| 3.2. Food | 52 | 62 | 38 | 43 |
| 3.3. Housing | 68 | 75 | 56 | 60 |
| 3.4. Consumer goods and services | 43 | 54 | 36 | 38 |
| **4. *Culture and spirituality*** * | 48 | 51 | 32 | 35 |
| 4.1. Level of services provided by cultural institutions | 32 | 38 | 22 | 29 |
| 4.2. Social compliance with high humanistic standards and values | 28 | 35 | 20 | 21 |
| 4.3. Kindness, assistance, and sympathy of surrounding population | 84 | 81 | 54 | 52 |
| **5. *Security in all aspects of life*** * | 31 | 33 | 21 | 21 |
| 5.1. International security and integrity of the State | 47 | 55 | 23 | 27 |
| 5.2. Right to private property and its protection | 46 | 45 | 36 | 33 |
| 5.3. Protection against poverty and unemployment | 17 | 16 | 13 | 13 |
| 5.4. Protection against crime | 15 | 17 | 12 | 13 |
| **6. *Work in chosen specialty*** * | 21 | 27 | 22 | 26 |
| 6.1. Work in chosen specialty | 24 | 33 | 24 | 31 |
| 6.2. Salary | 18 | 19 | 15 | 20 |
| **7. *Education and professionalism*** * | 43 | 45 | 29 | 29 |
| 7.1. Accessibility of professional education | 40 | 38 | 25 | 33 |
| 7.2. Education and professional level | 47 | 50 | 37 | 42 |
| 7.3. Level of political and economic knowledge | 53 | 47 | 30 | 29 |
| 7.4. Level of legal knowledge | 33 | 41 | 26 | 23 |
| 7.5. Level of environmental knowledge | 42 | 51 | 25 | 20 |
| **8. *Personal freedom and freedom of speech*** * | 55 | 61 | 31 | 31 |
| 8.1. Information on social situation | 61 | 68 | 38 | 39 |
| 8.2. Safeguards for rights and personal freedom | 56 | 55 | 28 | 26 |
| 8.3. Information on environmental conditions | 59 | 61 | 28 | 28 |
| **9. *Will to live*** * | 53 | 60 | 42 | 43 |
| 9.1. Ability to set daily goals and accomplish them | 56 | 61 | 48 | 58 |
| 9.2. Ability to adapt to new living conditions | 52 | 65 | 53 | 48 |
| 9.3. Belief that situation will improve | 51 | 53 | 25 | 23 |
| *Social satisfaction index* | 46 | 50 | 34 | 37 |

*Average of parameters in each category.

# 8
ЗКОНОМИЧЕСКИЙ

# ECONOMY
## *Chornobyl Accident Losses*

### V. Kholosha and V. Poyarkov

The Chornobyl Nuclear Power Plant (NPP) disaster was responsible for serious economic losses within the former Soviet Union and beyond. The accident disrupted production as well as the normal activities of daily life in many areas of Ukraine, Belarus, and Russian Federation. In Ukraine, it led to a significant loss of electrical power production and a direct impact on the regional industrial economy. Further, it caused substantial damage to the agricultural economy and limited the use of the area's forests and waterways (use restrictions were imposed on 5,120 km$^2$ of farmland and 4,920 km$^2$ of forest). For the entire Ukraine population, the reduction in the Gross National Product and the loss of monies that could have been spent on improved health care, medicine, and other areas to promote general health and well-being was a significant blow.

In 1986, approximately 116,000 persons were evacuated from areas with high radiation contamination. This evacuation required the construction of additional housing for the evacuees. Approximately 15,000 apartments; several living quarters with a total capacity exceeding 1,000 persons; 23,000 buildings; and approximately 800 social and cultural institutions were constructed during 1986 and 1987. The city of Slavutych was built to house former Pripyat residents (Chornobyl NPP workers and their families). Other people from the contaminated areas were located in Kiev.

The measures implemented by the authorities immediately following the accident were designed primarily to protect the public from the effects of radiation and minimize the immediate threat to human life and health. The evacuation was accompanied by various measures. To provide social and economic assistance to the public and individual enterprises, machinery, equipment, livestock, and other materials were relocated to less contaminated areas.

Assistance to the affected regions in Russia, Ukraine, and Belarus was provided from centralized all-Union financial and technical resources in the Soviet Union until 1991. The assistance focused primarily on restoring daily living activities. These activities included employment; restoring production activities (e.g., restarting evacuated industrial facilities, finding alternate power sources); decontaminating houses and roadways in areas believed to be salvageable; and providing social assistance, environmentally uncontaminated products, and medical services to members of the public who continued to reside in contaminated areas.

The 116,000 people who were evacuated from their homes were partially compensated for material losses related to the evacuation: lost personal property, crops in the ground, residences, etc. Industrial and agricultural enterprises (including collective farms) were compensated for lost financial, material, and technical resources.

In regions with radioactive contamination levels less than 555 kBq/m$^2$ (15 Ci/km$^2$), not subject to mandatory resettlement under the regulations, each resident was paid approximately 30 rubles ($33) per month to purchase uncontaminated food products imported from elsewhere. (At this time 1 kg of meat cost approximately 2 rubles and bread cost 20 kopecks.[1]) The use of local foodstocks (such as meat, milk, vegetables, and potatoes) was temporarily forbidden.

In 1990, the USSR Finance Ministry determined the direct losses as a result of the Chornobyl NPP accident. The losses were determined by analyzing data provided by various ministries and agencies, as well as the industrial departments of the USSR Council of Ministers and the councils of ministers of the union republics. The USSR Finance Ministry found that the total direct loss (including expenditures from all funding sources) for 1986–1989 was approximately 9.2 billion rubles or about 12.6 billion US dollars. As Ukraine's share of the all-Union budget was 30%, Ukrainian losses from the accident are in the same proportion.

In 1990, the USSR State Budget included 3.324 billion rubles for remediation of the Chornobyl NPP accident. Another 1 billion rubles was appropriated from the individual budgets of the Russian Federation and Ukrainian and Belorussian republics. The USSR State Budget for 1991 had included expenditures of 10.3 billion rubles for these purposes; however, because of the disintegration of the USSR, only a portion of the funding came from the all-Union budget. By the end of the year, remediation efforts were being

---

[1]Editor's Note: A kopeck is equal to one-hundredth of a ruble.

funded by the state budgets of the three newly independent and most severely affected countries (Russian Federation, Ukraine, and Belarus), Gosstrakh (an insurance company), and voluntary contributions to the Chornobyl NPP Accident Remediation Fund. A total of 2.97 million rubles of foreign currency resources (including 2.2 million dollars in convertible currency) were also received and used.[2]

## 8.1 ASSESSMENT OF DIRECT LOSS

### 8.1.1 Losses From the Shutdown of Economic Enterprises in Ukraine

After the Chornobyl disaster, several economic enterprises were no longer able to operate. These included

- The Dnepr River Fleet Repair and Operations Facility
- An iron foundry
- A cheese plant
- A vegetable drying plant (in the village of Novoshepelochi)
- A consumer services center.

The following organizations and institutions under local jurisdiction were located within the region:

- A fish cannery (Ivankovskii Rybokombinat—village of Ladizhichi)
- A regional agricultural equipment production association
- A regional agricultural chemicals production association
- A medical school
- An agricultural vocational and technical school
- 31 general education schools
- 49 medical institutions, including 5 hospitals and 15 recreation centers
- 39 clubs and movie theaters
- 43 libraries.

---

[2]Editor's Note: This information was officially presented at an ECOSOC (United Nations Economic and Social Council) meeting by the USSR, Belorussian, and Ukrainian delegations (letter No. A/45/342 and E/1990/102 dated 6 July 1990 addressed to the United Nations Secretary General).

The Chornobyl region included 64 population centers under the jurisdiction of City Soviets and 20 under the jurisdiction of Village Soviets. There were 374 km of roads, 353 km of which were paved. The 2,000-km² region (the inner Exclusion Zone) included the cities of Pripyat and Chornobyl, but not the villages of Dityatki, Gornostaypil, and Strakholissya. The region included 5,920 km² of farmland, of which 3,610 km² was arable; 16 collective farms and State farms used this land. See Table 8.1.1 for a listing of the agricultural enterprises in the Chornobyl NPP area. The area also included more than 900 km² of forest.

The city of Pripyat was completed in 1985 and had a population of 48,000 at the time of the accident. In 1986, the city contained three large enterprises under all-Union jurisdiction (Chornobyl NPP, the Jupiter plant, and an integrated residential construction plant); a vocational and technical school; a music school; a complex of hospital institutions; a recreation center; three libraries; and a movie theater.

The estimated losses from 1986–1996 resulting from loss of economic infrastructure (or facilities of economic value) in the 30-km Exclusion Zone are 1 billion rubles (Table 8.1.2).

**TABLE 8.1.1.** Agricultural Enterprises in the Chornobyl Nuclear Power Plant Region

| Name of Farm | Number of Population Centers on Farm | Area of Farm (amount of of arable land), hectares in thousands |
|---|---|---|
| Druzhba Collective Farm | 4 | 5.3 (3.5) |
| Zavety Lenina Collective Farm | 5 | 5.8 (3.5) |
| Imeni XX Partsezda Collective Farm | 6 | 4.5 (2.3) |
| Ukraina Collective Farm | 3 | 3.2 (1.7) |
| Dostizhenie Oktabrya Collective Farm | 4 | 1.4 (1.1) |
| Imeni Kalinina Collective Farm | 5 | 3.8 (1.6) |
| 1 Maya Collective Farm | 3 | 1.1 (0.8) |
| Kommunar Collective Farm | 4 | 3.2 (1.8) |
| Put k Kommunizmu Collective Farm | 3 | 1.6 (1.0) |
| Imeni Kirova Collective Farm | 3 | 2.4 (1.3) |
| Pobeda Collective Farm | 2 | 3.3 (2.0) |
| Krasnoe Polessye Collective Farm | 3 | 1.8 (1.0) |
| Rassvet Collective Farm | 3 | 2.2 (1.0) |
| Imeni Kuybysheva Collective Farm | 1 | 3.4 (1.7) |
| Pripyatskii State Farm | 7 | 11.8 (2.5) |
| Land for Komsomolets Poles'ya State Farm | 4 | 2.0 (1.2) |

**TABLE 8.1.2.** Losses from Ukraine Economic Facilities Removed from Service in the Exclusion Zone After the Accident

| Physical Facility Lost as a Result of Chornobyl NPP Disaster | Year of Valuation as Fixed Asset or Inventory Item | Cost of Fixed Assets or Inventory Items | |
|---|---|---|---|
| | | Rubles, thousands | Dollars, thousands |
| Facilities and expenses associated with stopping construction on ChNPP Phase III | 1986[a] | 99,028 | 136,120 |
| ChNPP Unit 4 | 1964[b] | 201,000 | 223,330 |
| Chornobyl-2 | 1984[c] | 97,700 | 137,027 |
| Enterprises in telecommunications equipment industry (1) | 1986 | 51,070 | 70,199 |
| Enterprises in the metallurgical industry (1) | 1986 | 44,700 | 61,443 |
| Enterprises in the construction materials industry (1) | 1986 | 7,750 | 10,653 |
| Enterprises in the river transportation industry (2) | 1986 | 21,050 | 28,935 |
| Paved roads (353 km) | 1986 | 60,550 | 83,230 |
| Enterprises in the woodworking industry (1) | 1986 | 4,720 | 6,488 |
| Enterprises in the concentrated feed industry (1) | 1986 | 4,550 | 6,254 |
| Enterprises for primary processing of agricultural raw materials (1) | 1986 | 4,900 | 6,735 |
| Enterprises in the food industry (1) | 1986 | 5,010 | 6,887 |
| Enterprises engaged in the repair of tractors and agricultural machinery (1) | 1986 | 760 | 1,045 |
| Enterprises in the forestry industry (1) | 1986 | 4,700 | 6,460 |
| Collective farms (14) | 1986 | 79,693 | 109,544 |
| State farms (2) | 1986 | 18,659 | 25,648 |
| Joint enterprises (3) | 1986 | 18,694 | 25,696 |
| Water systems and facilities | 1986 | 4,405 | 6,055 |
| Sewer systems and facilities | 1986 | 3,850 | 5,292 |
| Electrical transmission and distribution | 1986 | 315 | 433 |
| Heating systems and facilities | 1986 | 3,390 | 4,660 |
| Housing space:<br>• State-owned (402)<br>• Privately owned (2,278)<br>• Rural farmsteads (9,050) | 1986 | 209,750<br>7,101<br>28,200 | 288,316<br>9,761<br>40,005 |
| Vacation centers (10); hospital facilities (midwifery centers) (44); educational institutions in the vocational education system (3); general education schools (34); music schools (2); recreation centers (16); movie theaters (2); clubs (39) | 1986 | 29,104 | 40,005 |
| Total | | 1,010,649 | 1,338,979 |

(a) Exchange rate as of April 1986: $1 = 72.75 kopecks
(b) Exchange rate as of October 1984: $1 = 71.3 kopecks
(c) Exchange rate as of 1964  $1 = 90 kopecks
ChNPP = Chornobyl Nuclear Power Plant

### 8.1.2 Losses From Equipment Contamination During Accident Remediation

The substantial loss of infrastructure facilities in the Exclusion Zone was accompanied by further losses of equipment, tools, and machinery that became contaminated with radionuclides during accident remediation operations. These contaminated materials were disposed of at the Buryakovka Radioactive Waste Disposal Site and at the Rozsokha Equipment Holding Facilities 1 and 2. Items in the Buryakovka disposal site include 1,958 trucks, 14 fire trucks, and 19 bulldozers; the total estimated cost as of 1986 of the equipment in this disposal site was 17,566 thousand rubles or 24,146 thousand US dollars (estimated cost as of 1986). This is from internal accounting data from Kompleks State Enterprise. For more information on this site, see Section 4.1.8. Items in the Rozsokha holding facilities include 30 helicopters and 11 residential buildings; the total estimated cost as of 1986 of the equipment in this holding facility is 16 million rubles or about 22 million US dollars (estimated cost as of 1986). The total loss—loss of property and individual facilities of economic importance—was 1,044 million rubles or 1,385 million US dollars in the Exclusion Zone alone (estimated cost as of 1987).

## 8.2 ESTIMATE OF DIRECT COSTS

### 8.2.1 Direct Accident Remediation Costs

Accident remediation costs were estimated from the funding levels and cost breakdowns provided in the PO Kombinat and Pripyat organization annual reports for 1989 and 1990 (generally using interpolation of specific weights given to each type of work or service). Total expenditures from April 1986–October 1991 for Chornobyl NPP accident remediation efforts in the Exclusion Zone were 6,492 million rubles (or 8,923 million US dollars at the April 1986 exchange rate[3]). These expenditures covered activities in Table 8.2.1.

### 8.2.2 Social Protection and Corresponding Medical Programs

Measures for social protection of the victim population—primarily support and improvement of public health—have taken high priority in the funding of Chornobyl disaster remediation. Social protection includes special assistance, rehabilitation support, and physical fitness for individu-

---

[3]Editor's Note: The April 1986 exchange rate was 0.7275 kopecks to the dollar.

**TABLE 8.2.1.** Expenditures for Remediation in the Exclusion Zone
(April 1986–October 1991)

| Expenditure | Cost, rubles | Cost, US dollars at April 1986 exchange rate |
|---|---|---|
| *Activities on the Chornobyl NPP territory* | | |
| Construction, operation, and repair of the Shelter | 3.948 billion | 5.427 billion |
| Accident remediation and rehabilitation of Chornobyl NPP Units 1 - 3 | 1.674 billion | 575 million |
| *Specialized activities in Exclusion Zone* | | |
| Decontamination | 14.30 million | |
| Protection of water resources | 56.61 million | |
| Fire protection | 1.77 million | |
| Construction, renovation, and upgrade (includes expenses for conservation of facilities constructed and commissioned before the accident) | 842.53 million | |
| Radiation monitoring | 11.58 million | |
| Startup and adjustment | 8.09 million | |
| Scientific research | 407.84 million | |
| Experimental design | 331.63 million | |
| Total for specialized activities | 1.67 billion | 2.309 billion |
| *Measures to prevent the migration of radionuclides beyond the Exclusion Zone and Exclusion Zone access control* | | |
| Special measures implemented by Ministry of Internal Affairs units, including the Fire Protection Administration | 80.29 million | 110 million |
| Support of communications infrastructure | 15.37 million | 21 million |
| Support of road transportation | 59.12 million | 81 million |
| Support of air transportation | 5.22 million | 7 million |
| Support for heating, electricity, water service, and sewer service | 83.54 million | 115 million |
| Industrial, residential, and utility infrastructure | 4.90 million | 6.7 million |
| Medical and preventive dietary items and individual protective equipment | 177.38 million | 243.6 million |
| Maintenance and repair of Exclusion Zone roads and rail lines | 1.31 million | 1.8 million |
| Total for implementing special measures | 427.13 million | |
| Medical and rehabilitation insurance for remediation personnel, including Ukrainian Ministry of Health Public Health Organization annual physical examinations of personnel working in the Exclusion Zone, as well as physical fitness costs (free and subsidized passes to health facilities) | 24.9 million | 34.2 million |

als, especially children, affected by radiation. Expenditures related to social protection of the public are regulated by the Ukrainian Law on the Status and Social Protection of Citizen Victims of the Chornobyl Disaster. Such expenditures currently make up 60–65% of the total; in recent years, however, the limited funds available from the government have only been able to meet 33–50% of the need.

### 8.2.3  Scientific Research Programs

Scientific support has played a key role in resolving the problems related to the Chornobyl NPP accident. Applied scientific research and implementation of the results have been proceeding in the areas of medicine, radiobiology, radioecology, and environmental monitoring of radionuclides, radioactive waste decontamination and reprocessing, evaluation and prediction of regional environmental status, and social-psychological issues related to the Exclusion Zone and mandatory resettlement zone.[4] Fundamental research is being performed in radioecology, on the effect of low doses of ionizing radiation, and on radionuclide migration in soil, water, and air.

Such comprehensive research requires more extensive use of the existing research infrastructure and the creation of additional research infrastructure. The fundamental problem in addressing this need is the lack of adequate funds.

### 8.2.4  Radiation Monitoring of the Environment

The funds allocated to radiation monitoring of the environment (Table 8.2.2) have enabled researchers to do the following:

- Determine the radiation levels in contaminated areas, particularly in the voluntary resettlement zone
- Determine the detailed radionuclide distribution in areas where people continue to reside and work
- Measure radionuclide migration in the environment
- Conduct long-term observations in special test areas.

In addition, the program for studying the effects of radon on human health has played an important role in reducing the dose burden on Ukrainian

---

[4]Editor's Note: The Exclusion Zone is the area within 30 km of the Chornobyl NPP that was evacuated shortly after the accident. The mandatory resettlement zone is a highly radioactive area defined by Ukrainian Law on the Status and Social Protection of Citizen Victims of the Chornobyl Disaster in 1991. See Table 6.2.5 for more information.

**TABLE 8.2.2.** Expenditures for Accident Remediation and Social Protection of the Public for 1986–1996 (US dollars in millions)

| Line | Heading | 1986–1991 | 1992 | 1993 | 1994 | 1995 | 1996 | 1997[a] |
|------|---------|-----------|------|------|------|------|------|---------|
| 1 | Social protection of citizens, total | 6606.55 | 197.33 | 196.51 | 478.07 | 383.97 | 545.65 | 636,93 |
| 2 | Special assistance | 53.62 | 6.32 | 2.99 | 8.83 | 22.81 | 19.02 | 8,21 |
| 3 | Scientific research | 57.76 | 3.23 | 4.45 | 4.99 | 5.92 | 7.04 | 10,54 |
| 4 | Radiation monitoring | 63.79 | 1.99 | 1.64 | 2.28 | 3.15 | 4.44 | 5,4 |
| 5 | Environmental remediation | — | — | 0.01 | 0.37 | 0.36 | 0.19 | 0,23 |
| 6 | Rehabilitation and disposal of radioactive waste | 0.17 | 0.27 | 0.08 | 0.20 | 0.13 | 0.16 | 0,29 |
| 7 | Capital investment. Resettlement and creation of appropriate conditions for members of the public residing in contaminated areas | 31,73.62 | 276.07 | 197.78 | 205.28 | 167.44 | 194.10 | 89,87 |
| 8 | Work in Exclusion Zone | 8,923.75 | 19.70 | 25.84 | 46.45 | 44.95 | 52.08 | 56,1 |
| 9 | Other | 228.97 | 17.72 | 15.88 | 25.91 | 41.94 | 43.36 | 37,0 |
| | Total | 19,108.23 | | | | | | |
| | Ukrainian portion[b] | 5732,47 | 510.81 | 436.01 | 755.72 | 638.30 | 835.19 | 844,6 |

Note: The expenditures were funded by the USSR State Budget and Ukrainian SSR/Ukraine State Budget.

(a) An exchange rate of $1 = 1.86 hryvnia was used for 1997.
(b) Assuming expenditures were in the same proportion as for the overall operating portion of the all-Union budget (30%).

public (Section 6.3.2.3). Further, Ukraine has initiated development of the first "new generation" radiation monitoring instruments in support of health institutions, agricultural management authorities, forestry management authorities, management authorities for other sectors of the economy, and the public.

### 8.2.5 Decontamination and Radioactive Waste Management

Decontamination of land contaminated by the Chornobyl NPP disaster and burial of radioactive waste are intended to reduce the individual and collective dose to the public from these significant environmental sources. (For more information on the waste sites, see Chapter 4.) Concurrently with the decontamination operations, social and living conditions of the people residing in contaminated areas are being improved.

In contaminated regions occupied by a network of industrial and agricultural enterprises that are sources of anthropogenic pollutants, the negative effects of such pollutants on human health are amplified by the radiation factor. Implementation of various engineering and technical measures to eliminate the pollutants and limit their propagation will reduce the technologically induced hazard level. Special funds have been allocated for comprehensive scientific studies and analyses to assess and predict the environmental and health status of areas affected by radioactive contamination (Table 8.2.2). Actual progress is hampered by the lack of adequate funding.

## 8.3 ANALYSIS OF INDIRECT LOSSES

### 8.3.1 Losses from Inability to Use Contaminated Natural Resources

The land contaminated by the Chornobyl NPP accident in Ukraine includes rich forests where mushrooms and berries were harvested and agricultural lands where thousands of metric tons of hay were harvested. The loss of the ability to use farmland, water resources, and forest resources because of contamination is currently estimated to be 8.6–10.9 billion rubles. This is more than 2% of the gross national income produced by Ukraine in 1986. These figures are for Ukraine alone from 1986–1991. All economic activity was suspended on land with contamination densities greater than 555 kBq/m² (15 Ci/km²), and some activity was suspended on land with contamination densities between 185 kBq/m² and 555 kBq/m² (5 Ci/km² and 15 Ci/km²). It will take several decades for the contamination on this land to decrease sufficiently to permit use.

Forestry industries and fisheries also incurred significant losses. More than 5,000 km² of forest land was withdrawn from use. The direct losses due to loss of lumber were nearly 100 million rubles. The total loss incurred by forestry and related woodworking industries for the 1986–1991 time period was approximately 1.8–2.0 billion rubles (in 1984 prices).

Although only 0.6% of the pine stock in the former Soviet Union was located here, this area produced more than 50% of the total amount of resin collected in the former Soviet Union. Approximately 60,000 metric tons of coniferous sawdust per year, worth 15 million rubles, was collected here. The loss to water resources and fisheries in the Dnepr and Black Sea watersheds because of radioactive contamination in bodies of water during the first few years following the accident was 2.3–3.1 billion rubles.

### 8.3.2 **Loss of Electrical Production and its Industrial Impact**

Because of the accident, electrical power was not produced using the Chornobyl NPP and goods and services were not produced because of the loss of power. These losses are especially important relative to the other losses resulting from the Chornobyl NPP accident. The amount of electrical power not generated because Unit 4 was not used for its entire design life-time and because other Chornobyl NPP units were shut down in 1986 was 62 billion kWh. At a mean cost of 1.5 kopecks/kWh for Chornobyl NPP power, the direct loss was approximately 1 billion rubles. Economists esti-mate that each unit of electrical power cost supplied to other branches of industry increases national income by 20 units. Electrical power shortages have a substantial effect on production volume in areas such as machinery, light industry, food industry, and other processing industries. Thus, the total loss due to lack of electric power was approximately 20 billion rubles (in 1986 prices).

After the Chornobyl NPP accident, a moratorium was issued regarding bringing any new nuclear power plants on line at existing power plants. Because of this decision, the national economy failed to receive 6 mil-lion kW of installed capacity. Economists' estimates indicate that a mere 1-year delay in bringing 1 kW of electrical power on line is capable of reducing the national income by 2 billion rubles. If the delay becomes long term, the cost of the moratorium could reach 48 billion rubles (in 1984 prices) within 4 years.

Thus, to summarize the indirect losses, the total irretrievable loss to Ukrain-ian economy from the Chornobyl NPP disaster is 82–85.4 billion rubles (in 1984 prices). A breakdown of the indirect losses is provided in Table 8.3.1.

**TABLE 8.3.1.** Breakdown of Ukrainian Indirect Losses due to Chornobyl NPP Accident

| Indirect Loss | Rubles, billions |
|---|---|
| Loss of life and private property | 0.4–0.5 |
| Expropriation of agricultural land | 3.0–4.0 |
| Forestry industry losses | 1.8–2.0 |
| Water resources losses | 1.5–1.8 |
| Water resources and fisheries | 2.3–3.1 |
| Fixed assets | 5.0–6.0 |
| Cost of electricity not generated | 20.0 |
| Cost of moratorium against bringing new capacity on line at existing nuclear plants | 48.0 |
| Total | 82.0–85.4 |

## 8.4 ASSESSMENT OF TOTAL ECONOMIC LOSS DUE TO THE ACCIDENT

Historians and analysts believe that in the midst of the Reagan-era defense spending contest, the Chornobyl NPP accident was the straw that broke the economic back of the USSR. We, the authors, would like to emphasize that since its independence Ukraine has been independently funding accident remediation. Approximately 5–7% of Ukrainian State Budget has been spent on accident remediation, and this percentage has not decreased. Despite Ukraine's difficult economic situation, the percentage of the national budget devoted to the Chornobyl Fund shows no signs of decreasing within the next few years. This is primarily because of the provisions for social and medical assistance for the victims and issues involving the Exclusion Zone and the Shelter as described for the years 1997–2000 in the national program for minimizing the impact of the Chornobyl accident.

The total economic losses from the Chornobyl disaster are shown in Table 8.4.1.

**TABLE 8.4.1.** Breakdown of Total Ukrainian Economic Losses

| Item | Cost, million US dollars |
| --- | --- |
| Direct losses of inventories and economic assets | 1,385 |
| Direct expenditures for accident remediation funding during 1986 - 1991 | 5,732.5 |
| Ukrainian direct expenditures for accident remediation operations funding during 1992–1997 | 4,020.6 |
| Total: Direct losses and expenditures | 11,138.1 |
| Indirect losses (mean values) (Section 7.3) | 117,000 |
| Grand total | 128,138.1 |

These losses are not exhaustive, as they do not include all indirect Ukrainian economic losses but omit items such as

- Loss of health and fitness for work (for the current and future generations)

- Future costs for reclamation of contaminated land and bodies of water

- Future costs for disposal of radioactive waste from the Shelter or for the Shelter itself.

Under current law, the number of individuals requiring support as victims of the disaster is increasing. This is a back-breaking burden on Ukrainian economy (during a severe economic downturn) and reduces the overall effectiveness of expenditures on accident remediation. One potential way out of this problem would be to optimize the Chornobyl law so that it takes into account the real capabilities of the economy and sets priorities for accident remediation.

# 9
## окончнаннйе

# CONCLUSIONS
### *Lessons from the Accident*

V. Poyarkov, V. Bar'yakhtar, V. Kholosha, V. Kukhar',
V. Shestopalov, and I. Los'

The comprehensive assessment of the impact of the 1986 Chornobyl Nuclear Power Plant (NPP) accident presented here shows that there are three main risk categories:

- Risks to human health and the environment

- Subjective risks related to public perception of the accident's effects and the risk of economic loss

- Risks related to reexamination of general attitudes toward safety of nuclear power operations.

## 9.1 HUMAN HEALTH AND ENVIRONMENTAL RISKS

The accident at the Chornobyl NPP Unit 4 on 26 April 1986 was the most damaging nuclear accident in history. The explosion, fire, and spread of contamination presented and still present risks. The risks to human health and the environment are related to the following:

- Incomplete control and monitoring of high-level radioactive waste and nuclear material in the Shelter

- Incomplete monitoring and control of radioactive waste in Exclusion Zone storage and disposal sites

- High radionuclide contamination levels in the Exclusion Zone

- Radionuclide fallout outside the Exclusion Zone.

### 9.1.1 **Monitoring the Shelter**

As of 1 June 1996, the Shelter contained approximately $6 \times 10^{17}$ Bq of high-level radioactive waste. Radiation and other physical and chemical effects are transforming the radioactive waste. One product of the radioactive waste transformations is radioactive dust. A structural collapse of the Shelter could lead to the formation of a 100- to 200-m high plume containing more than 2,000 kg of dust and approximately 40 kg of finely dispersed fuel. Depending on meteorological conditions, this could significantly contaminate the area inside the Exclusion Zone and result in significant exposures to personnel on the site.

Another process related to radioactive waste transformation in the Shelter is uranium and plutonium leaching from the radioactive waste and accumulating in a specific location. Detailed prediction of this process is difficult, and although there is currently no danger of a critical mass forming, no one can guarantee that this will remain the case into the future.

The third hazard identified is the possibility of environmental contamination from migration of radioactive water in the Shelter (approximately 1,000 metric tons of radioactive water accumulates in the Shelter each year).

Shelter stabilization must be resolved, as partial or total collapse of the facility would significantly contaminate some area inside the 30-km Exclusion Zone, and the likelihood of this occurring continues to increase. Thus, stabilization of the Shelter and ensuring the nuclear materials inside the facility remain subcritical must be addressed immediately.

Two possibilities are under discussion for stabilizing the Shelter: entombing the waste within the Shelter or building a second shelter.

To entomb the waste within the Shelter, concrete or modified concrete containing neutron-absorbent admixtures (such as boron) would be poured on top of the radioactive waste in the Shelter. The concrete or admixture would limit the potential for redistribution of radionuclides in the Shelter, thereby preventing the onset of criticality. The main advantages of this plan are low cost and low personnel dose. The main disadvantage is that a century from now, there may be a desire to dispose of the remaining plutonium and other transuranic elements imprisoned in the concrete. However, this decision would shift some of the burden of converting the Shelter into a completely safe facility to future generations.

To build a second structure (Shelter 2), a facility that meets international standards would be constructed around the current facility. Then, the waste in the Shelter would need to be removed, reprocessed, and disposed of in the new shelter. The main advantage of this plan is the complete, timely resolution of the problem. The main disadvantages are the high cost and high personnel dose.

In 1996, Ukraine and the G-7 countries concluded an agreement to close the remaining operating reactors at the Chornobyl site and to render financial assistance to stabilize the Unit 4 Shelter and decommission the remaining reactors. Details of the agreement still have to be finalized; however, the likely solution is for a program to repair and transform the existing Shelter into an ecologically safe system.

For more information about the Shelter, see Chapter 3.

### 9.1.2 Monitoring Radioactive Storage and Disposal Sites in the Exclusion Zone

Conditions at the interim radioactive waste storage and disposal sites are not compliant with national or international recommendations for radioactive waste storage. Groundwater is an ongoing and significant concern. This means that the corresponding sources of radioactivity are not completely under control. However, analysis indicates that the collective dose received by Ukrainian public due to radionuclide migration from storage facilities is much smaller than the dose due to natural (background) radioactivity. On the other hand, lack of guaranteed control over the source under all possible conditions creates an additional risk, which should be minimized. Under current conditions, the most important task is to monitor access (i.e., prevent unauthorized access) to the storage facilities for several hundred years at a minimum (the biologically most important radionuclides have half-lives of 14.4 years ($^{241}$Pu) and 432 years ($^{241}$Am)).

For more information on the storage and disposal sites, see Chapter 4.

### 9.1.3 High Radionuclide Contamination Levels in the Exclusion Zone

In some areas of the Exclusion Zone, the $^{137}$Cs contamination levels reach 740,000 kBq/m$^2$ (20,000 Ci/km$^2$) or even higher. This means that the soil in these areas is basically equivalent to radioactive waste. Therefore, the natural ecosystems in these areas are exposed to high radiation levels. However, no significant radiobiological effects have been observed in these ecosystems to date. The uniqueness of this situation, in which all elements of an ecosystem are exposed to high radiation levels, will nevertheless undoubtedly require long-term monitoring of these systems, as well as scientific support for such monitoring.

Vertical and horizontal radionuclide migration will lead to localized contamination of groundwater. Migration processes over such large areas may have certain characteristics that have not yet been comprehensively studied. This may require continuous monitoring of radionuclide migration between the various interconnected elements of the environment, as well as the construction and verification of appropriate predictive models. At present, radionuclide transport from the Exclusion Zone does not make a substantial contribution to the collective dose received by Ukrainian public.

For more information on environmental contamination, see Chapter 5.

### 9.1.4 Radionuclide Fallout Outside the Exclusion Zone

The risks related to the effects of Chornobyl-related radiation on human health are a function of the individual effective dose and collective dose received. The individual doses are gradually decreasing each year in all of the areas contaminated by the accident. In virtually all population centers, the mean annual dose due to Chornobyl-related radiation is less than or equal to the mean annual dose received by a Ukrainian resident from natural sources (5 mSv/yr).[1] There are only 50 population centers where the mean annual individual dose is greater than 1 mSv/yr—the criterion for a population center to be considered part of the voluntary resettlement zone.

The collective dose received by Ukrainian public from accident-related radionuclides is a fraction of the collective dose from natural sources. Even in one of the most contaminated oblasts, Zhitomir Oblast, the dose from natural radionuclides is seven times larger than the Chornobyl-related dose. For the most contaminated areas, i.e., those in the mandatory resettlement zone (for example, the village of Denisovichi, Poliska Region, Kiyiv Oblast; approximate contamination density is 555 kBq/m$^2$), the dose from Chornobyl-related radionuclides would be approximately equal to that from natural radionuclides, if individuals lived there and consumed locally produced food.

External radiation only made a substantial contribution to collective dose during the first few years following the accident. As of 1998, more than 70% of the Chornobyl-related collective dose is due to internal exposure, with $^{137}$Cs playing the dominant role. The internal exposure is almost entirely due to food products, whose $^{137}$Cs content depends on the soil-to-vegetation transfer coefficient. Research indicates that the $^{137}$Cs uptake in agricultural products decreases from one year to the next. Thus, the Chornobyl-related component of the collective exposure is expected to gradually decrease with time.

As of 1998, 978 of the 2,293 population centers legally classified as being within the guaranteed voluntary resettlement zone now have estimated exposure levels well below the criteria used to determine their classification.

---

[1]Editor's Note: In the United States, this value is assumed to be 3.6 mSv/yr.

Thus, based on the direct exposure to Chornobyl-related radionuclides (neglecting social and psychological effects), the current law protecting the residents of contaminated areas is probably excessively conservative.

For more information on radiation risks to the public and accident remediation workers, see Chapter 6.

## 9.2 PUBLIC PERCEPTION OF THE ACCIDENT EFFECTS AND ECONOMIC LOSS

### 9.2.1 Risks and Public Perception of the Accident's Effects

Social risks, particularly the subjective component of these risks, is an important part of understanding the impact of the Chornobyl NPP accident. This problem is important not only for understanding the post-Chornobyl behavior of victims, but also for analyzing public behavior relative to phenomena such as smoking where there is clearly a relationship between involuntary risk, voluntary risk, objective risk, and subjective risk. This research has been only a first attempt at understanding how this new field of study applies to the Chornobyl accident. Clearly this area deserves further study.

For more information on the social impacts of the accident, see Chapter 7.

### 9.2.2 Risk to Personal Health due to Economic Damage from the Accident

It is well known that public health in any country is 80–85% determined by the economy of that country, with health being defined as total physical, psychological, and spiritual well-being, and not just the absence of disease.

The Chornobyl NPP accident undoubtedly caused the country to suffer an enormous loss from the following:

- Direct loss of assets and facilities of economic value

- Electrical power not generated and other indirect costs

- Economic costs to minimize the effects of the accident in lieu of expenditures on social needs, sanitation, or other socially useful purposes.

The total expenses and losses due to the Chornobyl disaster are estimated to be approximately 128 billion US dollars, or 2.5 times Ukraine gross domestic product for 1997. These funds could have been used during the same period for economic development, science, or other activities that would have improved the wealth of the people and country.

For more information on economic losses, see Chapter 8.

## 9.3 REEXAMINATION OF ATTITUDES TOWARD SAFETY OF NUCLEAR POWER OPERATIONS

Most modern technology has some risks to life and health, but humanity has made the choice to build a technocratic civilization. And, the positive results are generally apparent—the mean lifespan has increased by a substantial factor in all civilized countries. However, sources of technological risk require increased attention, and this is a necessary attribute of modern life. The experience gained from the Chornobyl NPP accident is of enormous value for humanity to understand the need to develop risk management systems for hazardous facilities.

### 9.3.1 Regulatory Infrastructure and Reactor Design Improved in Ukraine

Analyses of the root causes of the accident indicate that the regulatory infrastructure of that time was insufficient to guarantee nuclear safety. This was taken into account, and Ukraine established a regulatory authority for nuclear and radiation safety separate from the manufacturers and operator— the State Committee for Nuclear and Radiation Safety, which later became the Ministry of Environmental Protection and Nuclear Safety.

In addition to the regulatory infrastructure, the safety system within the initial design of the RBMK reactors did not meet current safety standards. With the exception of the lack of adequate containment, the design flaws that led to the accident were largely removed between 1987 and 1991. Multiple safety barriers, emergency protection systems, and other control and monitoring systems were installed to minimize the risks at the facility level. Also, personnel training, nondestructive assessment, and safety analysis were improved. As a result, it seems impossible for the same type of accident to occur again. However, other accidents leading to the release of substantial amounts of radioactive material cannot be ruled out [IAEA 1996a].

One reactor design area that still needs improvement is the emergency warning system. The system has not been integrated with standard response procedures, measurement procedures, environmental contamination determination procedures, prediction methods, and decision-making rules at the Chornobyl NPP.

### 9.3.2 Emergency Planning Needs to be Improved

The experience gained through the Chornobyl NPP accident will be invaluable in future emergency planning and response work. Emergency planning and response is based on averting acute (4–6 h), short-term (a week), and long-term doses (over a 50- or 70-year time interval) [Poyarkov 1997].

A critical period in handling a nuclear emergency is the acute radiation assessment period. During this period, scientists predict the areas where the dose may exceed 1 Sv during the release or the 4–6 h following the release, thereby providing a basis for protection against doses that would produce deterministic effects. The main objective during the short-term radiation assessment is to use emergency radiological monitoring data as well as the measures already in place to minimize the release and protect personnel and the public, to predict the areas where the dose will exceed 1 Sv during the next month. All countermeasures, implemented and planned, are taken into account in this assessment. The main goal during the long-term radiation assessment period is to predict the area where the dose over a 50- or 70-year period might exceed maximum permissible levels (allowing for all implemented countermeasures). The two latter emergency response stages require short- and long-term assessment of the corresponding doses using radiological measurements and computer-based models.

An enormous amount of experience was obtained in post-accident monitoring as a result of the Chornobyl remediation effort; this experience could serve as the cornerstone of a scientific decision-making support system for nuclear and radiological accidents. Development of such a system, based on a combination of accident assessment and monitoring procedures at the regional and international levels, is a high-priority task. This will be a good example where the Chornobyl NPP disaster, which brought so much suffering and misfortune, could also yield a positive social contribution.

A comprehensive system should be developed to create sophisticated national and international approaches to minimize the objective and subjective risks associated with severe accidents. The elements of this system are as follows:

- Education of government officials, legislators, the media, university students, school students, and the public concerning various types of accidents and disasters, as well as techniques and equipment for collective and personal protection

- Development of a decision-making system and models of optimum actions to be taken by the governmental authorities, legislators, and various agencies, services, governmental and nongovernmental organizations, and the media during the pre-accident phase, the accident itself, and the post-accident phase.

- Special studies of the laws governing subjective public perception of the effects of radiation exposure to influence this process and minimize the negative health effects of the process.

## 9.4 CONCLUSIONS

The huge experience accumulated during the cleanup of the Chornobyl accident consequences should be used as the basis for the scientific support of the decision making in the case of nuclear and other radiological accidents. The creation of such a system, that is based on the procedures of the estimations and monitoring of the accidents that are unified at the regional and international levels, is an important task. It would be a good example of how the Chornobyl catastrophe, which has brought much distress and suffering, will give a positive contribution to the development of the society.

# REFERENCES

Anonymous. 1988. *Nuclear Engineering and Design*, 106(N2):179.

Anonymous. 1994. *Technological process regulations for the Shelter Facility/ChNPP Reactor No. 4.* No. 1R-OU, Chornobyl Nuclear Power Plant, Chornobyl.

Abagyan, A.A., et al. 1986. "Information on the accident at Chornobyl NPP and its effects as prepared for the IAEA." *Atomnaya Energiya* 61(5):301-320.

Adamov, V.E., V.P. Vasil'evskii, and A.I. Ionov. 1986. "Analysis of the initial phase in the development of the accident in Chornobyl NPP Unit 4." *Atomnaya Energiya* 64(1):24-28.

Afonin, S.V., Vasil'chenko, Yu.P. Ivanov, et al. 1992. "Radiation monitoring and protection of bodies of water within the Chornobyl NPP 30-km zone." In *Papers from the Third All-Union Science and Engineering Conference on the Cleanup at Chornobyl NPP*, Vol. 1, part 1, pp. 221-230. Zelenyi Mys.

Anderson, E.B., S.A. Bogatov, A.A. Borovoi, et al. 1993. *Lava-like fuel-containing material in Shelter Facility.* Ukrainian Academy of Sciences Preprint 93-17, Shelter Facility Interindustry Science and Technology Center, Kiev.

Arkhipov, A.N. 1995. "Behavior of $^{90}$Sr and $^{137}$Cs in agricultural ecosystems within the Chornobyl NPP Exclusion Zone." PhD dissertation (abstract), Chornobyl.

Arkhipov, A.N., N.P. Arkhipov, D.V. Gorodetskii, and G.S. Meshalkin. 1994. *Development of the radioenvironmental situation on farmland in the Chornobyl NPP 30-km zone.* NPO Pripyat' preprint, Kiev.

Arkhipov, N.P. 1994. *Role of natural and man-made factors in radionuclide migration through soil and vegetation in various areas.* PhD dissertation (submitted), Obninsk.

Askew, J.R., et al. 1966. "A general description of lattice code WIMS." *J. BNES*, Vol. 5 (October).

Atimizdat. 1976. *Nuclear safety regulations for nuclear power plants*. PBYa-04-74, Atimizdat, Moscow.

Atomizdat. 1974. *General safety provisions for nuclear power plants during design, construction, and operation*. OPB-73, Atomizdat, Moscow.

Atomizdat. 1984. *General safety provisions for nuclear plants during design, construction, and operation*. OPB-82, Atomizdat, Moscow.

Begichev, S.N., A.A. Borovoy, Ye. V. Burlakov, et al. 1990. *ChNPP Unit No. 4 reactor fuel (a short handbook)*. Preprint 5268/3, Institute of Atomic Energy, Moscow.

Belli M., A. Marchetti, Yu. Ivanov, et al. 1994. "Radionuclides behavior in meadows." In *1993-1994 Final Report—ECP5 Project The Behavior of Radionuclides in Natural and Semi-Natural Environments*, pp. 73-84. ANPA, Rome.

Belyaev, S.T. and A.A. Borovoi. 1991. "Radioactive releases from Chornobyl NPP Unit 4." Presented at Nuclear Accidents and the Future of the Power Industry. 15-17 April 1991, Paris.

Belyaev, S.T., A.A. Borovoi, V.G. Volkov, et al. 1990. *Technical substantiation of Shelter Facility nuclear safety*. Kurchatov Institute Field Station, Chornobyl.

Bogatov, S.A. and S.L. Gavrilov. 1996. *Analysis of Current Shelter Facility Safety and Predictions for Development of Situation*. Kurchatov Institute Report No. 245NT-96, Kurchatov Institute, Moscow-Chornobyl (21 March).

Bogatov, S.A., A.A. Borovoi, V.I. Dvoretskii, et al. 1990. *Stability of the radiologically most hazardous radionuclides in various forms of the fuel release from the Chornobyl accident*. Preprint 5022/3, Kurchatov Institute, Moscow.

Bogatov, S.A., A.A. Borovoy, A.N. Kiselev, et al. 1991. *Estimate of the erosion rate of fuel-containing materials inside the Sarcophagus and a characterization of the resulting particles*. Preprint IAE-5434/3, Moscow.

Bogatov, S.A., A.A. Borovoy, A.S. Yevstratenko, et al. 1994. "Nuclear safety of the Shelter Facility." Chapter in *Technical Substantiation of Nuclear Safety*. No. 09-13-51. Shelter Facility Interindustry Science and Technology Center, Chornobyl (October 24).

Bogatov, S.A., A.A. Borovoy, S.A., Gavrilov, S.A. et al. 1995. *Nuclear, radiation and environmental safety of the shelter facility (collection, verification and presentation of information; additional research)*. Kurchatov Institute, Moscow.

Bondar', P.F., Yu.A. Ivanov, V.G. Avdeev, and A.G. Ozornov. 1989. "Estimated relative accessibility of radionuclides of unknown form following deposition in soil." In *Abstracts of Papers from the First All-Union Radiobiological Conference*, Vol. 5, pp. 1176-1177. 21-27 August 1989, Moscow.

Bondar', P.F., N.A. Loshchilov, A.I. Dutov, A.G. Ozernov, N.L. Sivendyuk, N.R. Tereshchenko, and A.V. Maslo. 1991. "General laws governing contamination of plant products in radioactively-contaminated areas." *Problemy Sel'skokhozyaistvennoi Radiologii* (1):88-105.

Bondar', P.F., Yu.A. Ivanov, B.S. Prister, G.P. Perepelyatnikov, L.V. Perepelyatnikova, and A.I. Dutov. 1996. "Agricultural aspects in the assessment and improvement of the radiological environment in contaminated areas." In *Abstracts from the Fourth International Scientific and Technical Conference—Chornobyl-94: Summary of Eight Years of Accident Cleanup at ChNPP*, Vol. 1, pp. 309-323. Zelenyi Mys.

Bondarenko, G.N. 1994."Geochemical migration characteristics of radionuclides from the Chornobyl release." In *Abstracts from the Fourth International Scientific and Technical Conference—Chornobyl-94: Summary of Eight Years of Accident Cleanup at ChNPP*. pp. 66-67. Zelenyi Mys.

Borovoi A.A. 1989. "Fission product and transuranic release during Chernobyl accident." In *materials of international conference: The fission of nuclei—50 years.* Leningrad.

Borovoi A.A. 1990. *Analytical report (post-accident management of destroyed fuel from Chernobyl).* International Atomic Energy Agency, Work Material, pp.1-99.

Borovoi, A.A., B.Ya. Galkin, A.P. Krinitsyn, et al. 1990a. "Newly-formed Fuel/Structural Material Interaction Products From ChNPP Unit 4. Reports 1 and 2." *Radiokhimiya* 32(6):103-113.

Borovoi, A.A., G.D. Ibragimov, O.O. Ogorodnikov, et al. 1990b. *Condition of ChNPP Unit 4 and nuclear fuel contained therein.* Kurchatov Institute Field Station Preprint. Chornobyl.

Borovoi A.A., A.R. Sich, and G.A. Dunbar. 1994. "Problems associated with the Chernobyl Unit-4 'Sarcophagus.'" In *The 1994 international symposium on decontamination and decommissioning*, April 25-28. Knoxville, Tennessee.

Borovoy, A.A., A.A. Dovbenko, M.V. Smolyankina, and A.A. Stroganov. 1991a. *Determination of the nuclear physics properties of ChNPP Unit No. 4 fuel.* IBRAE Report No. 52/11-20, USSR Academy of Sciences, Moscow.

Borovoy, A.A., B.Ya Galkin, L.V. Drapchinskiy, et al. 1991b. "Newly-formed fuel/structural material interaction products from ChNPP Unit 4. Report 3." *Radiokhimiya* 33(4): 177-196 (1991).

Borsilov V.A. 1996. "Physico-mathematical modeling of radionuclide behavior." *Radiation protection dosimetry* 64(1/2):3-11.

Brodkin, E.B. 1993. *Calculation and refinement of calculation critical parameters for various FCM types at locations in the Shelter Facility.* Stage 3 Report, Kurchatov Institute.

Bunzl K., P. Jacob P., Yu. Ivanov, et al. 1994. "Migration Behavior of Radiocesium in Meadows and External Radiation Exposure." In *1993-1994 Final Report- ECP5 Project, The Behavior of Radionuclides in Natural and Semi-Natural Environments*, pp. 85-107. ANPA, Rome.

Buzunov, N., Omelyanetz, N. Strapko, et al. 1986. "Chernobyl NPP consequences cleaning up participants in Ukraine health status epidemiologic study main results, The radiological consequences of the Chernobyl accident." In *Proceedings of the first international conference*. 18-22 March 1996. Minsk, Belarus.

*Byulleten Ekologicheskaya Sostoyanie Zony Otchuzhdeniya*, No. 4.

*Byulleten Ekologicheskaya Sostoyanie Zony Otchuzhdeniya*, No. 5.

Chumak, V.V., I.A. Likhtarev, and V.S. Repin. 1989. "Formalized maps and self-interviews for reconstruction of internal and external doses and related software/algorithms." In A*bstracts of papers presented at the all-union conference on current issues in dosimetry of internal radiation*, pp. 23-24. Moscow.

Demchuk V.V., O.V. Voitsekhovich, V.A. Kasparov, et al. 1990. "Analysis of Chernobyl fuel particles and their migration characteristics in water and soil." In *Proceedings of Seminar on Comparative Assessment of the Environmental Impact of Radionuclides Released during Three Major Nuclear Accidents: Kyshtym, Windscale, Chornobyl*. 1-5 October 1990, Luxembourg. Vol. 2, Commission of the European Communities, Radiation Protection-53, EUR 13574, 1991, pp. 493-513.

Dushin, V.N., B.F. Petrov, L.A. Pleskachevskiy, et al. 1992. "Confinement of intense gamma emitters and assessment of fuel quantities in the central hall of Unit No. 4." Radium Institute Scientific Production Association, Chornobyl.

Eisenbud, M. and T. Gessel. 1991. *Environmental radioactivity—from natural, industrial, and military sources*. 4th edition, Academic Press, San Diego, California.

European Union. 1996. *Atlas of cesium deposition on Europe after the Chernobyl accident*. Report 16733, EUR-1996, Office of Publication, Luxembourg.

Firsakova, S.K., N.V. Grebenshchikova, A.A. Novik, S.F. Timofeev, G.I. Palekshanova, and N.I. Samuseva. 1990. "Radionuclide migration in meadows contaminated by the accident at ChNPP." In *Chornobyl-90: papers from the second international scientific and technical conference on Chornobyl NPP accident cleanup results*, Vol. 2, Part 1, pp. 93-99. Chornobyl.

Gerasimova, T.S., I.Ya. Simanovskaya, and A.I. Surin. 1992. *Preliminary site selection for labeling and selection of testpoints for water sampling at the Shelter Facility*. 09/47, Nuclear and Radiation Safety Department, Shelter Facility, Interindustry Science and Technology Center, Chornobyl (27 December).

Gmal B. 1994. *Kritikalitatsabschatzungen zum Zustand des Sarkophags von Tschernobyl*. GRS, Germay (12 August).

Goulko G.M., V.V. Chumak V.V., N.I. Chepurny, et al. 1996. "Estimation of [131]I thyroid doses for the evacuees from Pripjat." *Radiation Environmental Biophysics* N35:81-87.

Governmental Commission on Comprehensive Resolution of ChNPP Issues. 1997. *Strategy for Conversion of the Shelter Facility into an Environmentally Safe System. Decision No. 5 of the Governmental Commission on Comprehensive Resolution of ChNPP Issues, 18 April 1997.*

Green B.M.R., J.S. Hughes, and P.R. Lomas. 1991. *Radiation atlas: natural sources of ionizing radiation in Europe.* NRPB, Chilton, Didcot, Oxon, United Kingdom.

Gudzenko, V.V., S.P. Dzhepo, D.A. Bugai, and A.S. Skal'skii. 1994. "Determination of distribution coefficients for radionuclides in the water-rock skeleton system." *Problemy Chornobil'skoi Zony vidchuzhdennya* (1), 93-96.

Illichev, S. V., O.A. Kochetkov, V.P. Kryuchkov, V.K. Mazurik, A.V. Nosovskii, D.A. Pavlov, I.V. Snisar, and A.G. Tsoy'yanov. 1996. *Retrospective dosimetry of accident remediation personnel at the Chernobyl Nuclear Power Plant.* ed A.V. Nosovskii, Seda-Stil, Kiev.

International Atomic Energy Agency (IAEA). 1992. *International Chernobyl project, technical report.* International Atomic Energy Agency, Vienna, Austria.

International Atomic Energy Agency (IAEA). 1996a. *Brief presentation of results from the decade since Chornobyl conference.* IAEA INFCIRC 510, International Atomic Energy Agency, Vienna, Austria.

International Atomic Energy Agency (IAEA). 1996b. *A decade after Chornobyl: environmental impact and future prospects.* IAEA/J1-CN-63, International Atomic Energy Agency, Vienna, Austria.

International Commission on Radiological Protection (ICRP). 1959. *Report of Committee II on permissible dose for internal radiation.* ICRP Publication 2, Pergamon Press, Oxford.

International Commission on Radiological Protection (ICRP). 1979. *Recommendations of the International Commission on Radiological Protection.* Publication 30, Pergamon Press, New York.

International Commission on Radiological Protection (ICRP). 1991. *Recommendations of the International Commission on Radiological Protection.* Publication 60, Pergamon Press, New York.

Ivanov, Yu.A. and V.A. Kashparov. 1992. "Behavior of radionuclides contained in the fuel component of ChNPP fallout in soil." *Radiokhimiya* 5:112-124.

Ivanov, Yu.A., V.A. Kashparov, L.A. Oreshich, P.F. Bondar', O.V. Hecheporenko, and G.I. Agapkina. 1991. "Physical and chemical forms of fallout from ChNPP release and their transformation in soil." In *Materials from the all-union conference on radioenvironmental and economic and legal aspects of land use after the accident at Chornobyl NPP*, pp. 234-239. Kiev.

Ivanov, Yu.A., N.A. Loshchilov, L.A. Oreshich, V.A. Kashparov, and P.F. Bondar'. 1992. "Soil dynamics of mobile forms of cesium-137 fallout from the ChNPP accidental release." *Problemy Sel'skokhozyaistvennoi Radiologii* (2):43-56.

Ivanov Yu., V. Kashparov, J. Sandalls, G. Laptev, N. Victorova, S. Kruglov, B. Salbu, D. Oughton, and N. Arkhipov. 1996a. "Fuel component of ChNPP release fallout: properties and behavior in the environment. The radiological consequences of the Chernobyl Accident." In *Proceedings of the first international conference*, EUR

16544 EN, eds. A. Karaoglu, G. Desmet, G.N. Kelly, H.G. Menzel, pp.173-177. 18 to 22 March 1996, Minsk, Belarus. Luxembourg.

Ivanov, Yu.A., V.A. Kashparov, S.I. Zvarich, and S.E. Levchuk. 1996b. "Vertical transport of ChNPP-released radionuclides in soil. II. Experimental modeling of vertical radionuclide transport in soil." *Radiokhimiya* 38(3), 272-277.

Izrael, Yu.A. 1984. *Ecology and environmental monitoring*. Gidrometeoizdat, Moscow.

Izrael, Yu.A., ed. 1990. *Chornobyl: radioactive contamination of the environment*, Gidrometeoizdat, Leningrad.

Izrael, Yu.A., et al. 1987. "Radioactive contamination of the environment in the vicinity of the Chornobyl Nuclear Power Plant." *Meteorologiya Gidrologiya* (2):5-18.

Kaplan, S., and B.J. Garrick. 1987. "On the quantitative definition of risk." *Risk Analysis* 1(1):1-27.

Karavaeva, E.N., N.V. Kulikov, I.V. Molchanova, V.N. Pozolotin, and P.I. Yushkov. 1990. "Migration and biological effects of radionuclides in the ChNPP forest landscape." In *Papers from the first international conference on the biological and radioenvironmental aspects of the impact of the Chornobyl NPP accident*, Vol. 1 (*Radioecology of vegetation. radioecology of land animals. radioecology of aquatic life*), ed. E.V. Senina, pp. 12-26. Zelenyi Mys.

Kashparov V.A., Yu.A. Ivanov, S.I. Zvarich, et al. 1996. "Formation of hot particles during the Chernobyl Nuclear Power Plant accident." *Nuclear Technology* 114:N.1 (May).

Kashparov, V.A., et al. 1997a. "Determination of maximum effective temperature and non-isothermal annealing time for Chornobyl fuel particles during the accident." *Radiokhimiya* 39(1).

Kashparov, V.A., Yu.A. Ivanov, S.I. Zvarich, V.P. Protsak, Yu.V. Khomutinin, and E.M. Pazukhin. 1997b. "Determination of solubility rates for Chornobyl fuel particles under natural conditions." *Radiokhimiya* 39(1):71-76.

Kazakov, V., and V.I. Berchii. 1994. "Ecological condition of the Exclusion Zone." Presented at Session of the Science and Technology Committee of the Administration of the Exclusion Zone, Chornobyl.

Khlopin Radium Institute. 1989. *Analysis of fuel/structural material interaction at the Shelter Facility*. Final Report Under Contract 2-94/89 N11.07/16. Khlopin Radium Institute, St. Petersburg (12 December).

Khlopin Radium Institute. 1991. *Analysis of fuel/structural material interaction at the Shelter Facility*. Final Report Under Contract 39-901/63-16-2-1, N11.07/285. Khlopin Radium Institute, St. Petersburg (13 December).

Khlopin Radium Institute. 1992. *Analysis of the Shelter Facility FCMs for identification of neutron absorber microimpurities.* Reference Report Under Contract 04-92/62, No. 09/35. Khlopin Radium Institute, St. Petersburg (21 December).

Kirchner G., and C. Noack. 1988. "Core history and nuclide inventory of Chernobyl core at the time of accident." *Nucl. Safety* 29(1):1-5. (January-March).

Konoplev A.V., and A.A. Bulgakov. 1992. "Behavior of Chernobyl-origin hot particles in the environment." In *Proceedings of the intern. symposium on radioecology. chemical speciation—hot particles,* Znojmo.

Konoplev, A.V., and A.V. Golubenkov. 1991. "Modeling the vertical migration of radionuclides in soil (using the results of a nuclear accident)." *Meteorologiya Gidrologiya* (10):62-68.

Konoplev A.V., V.A. Borzilov, A.A. Bulgakov, A.I. Nikitin, M.A. Novitsky, and O.V. Voicehovitch. 1992. "Case study no 8. hydrological aspects of the radioactive contamination of water bodies following the Chernobyl accident." In *Hydrological aspects of accidental pollution of water bodies.* pp. 167-190. WMO-No 754, Secretariat of the World Meteorological Organisation, Geneva, Switzerland.

Krisyuk, E.M. 1989. *Indoor radiation background,* Energoatomizdat, Moscow.

Krivokhatskii, A.S., E.A. Smirnova, Yu.V. Rogozin, V.A. Avdeev, and A.A. Kuskov. 1990a. "Radionuclide forms in surface samples and deep samples of soil collected from the Chornobyl 30-km zone between 1986 and 1988." In *Chornobyl-90: papers from the second international scientific and technical conference on Chornobyl NPP accident cleanup results,* Vol 2, Part 1, pp. 62-71. Chornobyl.

Krivokhatskii, A.S., E.A. Smirnova, V.G. Savonenkov, V.A. Avdeev, and A.A. Kuskov. 1990b. "Leachage of radionuclides from soil and man-made particles in the West Plume from the accident at ChNPP." In *Chornobyl-90: papers from the second international scientific and technical conference on Chornobyl NPP accident cleanup results,* Vol 2, Part 1, pp. 62-71. Chornobyl.

Kruglov, S.V., and N.P. Arkhipov. 1996. "Behavioral of $^{137}$Cs and $^{90}$Sr from Chornobyl fallout in soil from areas at large and small distances from the accident." In *Chornobyl-96: abstracts from the fifth international scientific and technical conference on the results obtained in 10 years of Chornobyl NPP accident cleanup* pp. 230-231. Zelenyi Mys.

Kruglov, S.V., A.D. Kurinov, and N.P. Arkhipov. 1996. "Forms of radionuclides in the soil within the Chornobyl 30-km zone and time variation in these forms of radionuclides." In *Abstracts from the fourth international scientific and technical conference—Chornobyl-94: summary of eight years of accident cleanup at ChNPP,* Vol. 1, pp. 243-250. Zelenyi.

Kryuchkov, V.P. and A.V. Nossovskii, eds. 1996. *Retrospective dosimetry of accident remediation personnel at the Chernobyl Nuclear Power Plant.* SEDA-STIL, Kyiv.

Kulakovskaya, T.N. 1965. *Agrochemical properties of soils and their significance for the application of fertilizer*. Urozhai, Minsk.

Kuriny V.D., Yu.A. Ivanov, V.A. Kashparov, et al. 1993a. "Particle associated Chernobyl fall-out in the local and intermediate zones." *Annals of Nuclear Energy* 20(N.6):415-420.

Kurnosov, V.A., V.M. Bagryanskiy, et. al. 1988. "Disposal of Chornobyl NPP Unit No. 4." *Atomnaya Energiya* 64(4):248.

Ledenev, A. P. Ovcharenko, I. Mishushko, V. Antropov. 1995. *Results of complex research of radioactive storage sites in the Exclusion Zone. Problems of Chernobyl Exclusion Zone, #2*, p. 46.

Likhtarev I., Kovgan L., Bobilova O. et al. 1994a. *Main Problems in Post-Chernobyl Dosimetry. Assessment of the Health and Environmental Impact from Radiation Doses due to Released Radionuclides*, Proceedings of the International Workshop at Chiba, January 18-20, 1994, pp. 27-51.

Likhtarev I.A., et al. 1994b. "Thyroid dose assessment for the Chernigov Region (Ukraine): estimation based on [131]I thyroid measurements and extrapolation of the results to districts without monitoring." *Radiation and Environmental Biophysics* (33):149-166.

Likhtarev, I.A. L.N. Kovgan, S.E. Vavilov, R.R. Gluvchinsky, O.N. Perevoznikov, L.N. Litvinets, L.R. Anspaugh, J.R. Kercher, and A. Bouville. 1996a. "Internal exposure from the ingestion of foods contaminated by the [137]Cs after the Chernobyl accident, Report 1. General model: ingestion doses and countermeasures effectiveness for the adults of Rovno oblast of Ukraine." *Health Physics* 70(3).

Likhtarev, I.A., L.N. Kovgan, S.E. Vavilov, D. Novak, P. Jacob, and H.G. Paretzke. 1996b. "Effective doses due to external irradiation from the Chernobyl accident for different population groups of Ukraine." *Health Physics* 70(1).

Los', I.P. 1993. "Health assessment of dose-producing ionizing radiation sources of natural and manmade origin and exposure doses received by the population of Ukraine." PhD dissertation (abstract), Kiev University, Kiev.

Loshchilov, N.A., V.A. Kashparov, Ye.B. Yudin, et al. 1991. "Experimental assessment of radioactive fallout from the Chernobyl accident." *Sicurezza e Protezione* N 25-26.

Loshchilov, N.A., V.A. Kashparov, V.D. Polakov, et al. 1993a. "Nuclear-physics characteristics of hot particles formed by the Chernobyl NPP accident." *Soviet-Radiochemistry, USA* 24(4):510-514 (March).

Loshchilov, N.A., V. A. Kashparov, Ye.B. Yudin, and V.P. Protsak. 1993b. "Fractionation of radionuclides in Chernobyl hot fuel particles." *Soviet-Radiochemistry, USA* 34(5):617-623 (May).

Lukashev, K.I., ed. 1961. *Geochemical characteristics of lithogenesis and landscape in the Belarus Poles'e*. Nauka i Tekhnika, Minsk.

Lukashev, K.I., and I.N. Petukhova. 1965. "Soil and geochemical provinces of the Belorussian Poles'e." *Dokl. Akad. Nauk Beloruss. SSR* 7(9):468-472.

Makhon'ko, K.P., ed. 1990. *Handbook for setting up environmental monitoring near nuclear power plants.* Gidrometeoizdat, Leningrad.

Makhon'ko, K.P., F.A. Rabotnova, and A.A. Volokitin. 1993. "Distribution of $^{137}$Cs produced in the ChNPP accident over USSR territory." In *Radiation aspects of the Chornobyl Accident*, Vol. 1 (*Radioactive Contamination of the Environment*), pp. 252-259. Gidrometeoizdat, Moscow.

Marei, A.N., A.S. Zykova, and M.M. Saurov. 1984. *Communal radiation hygiene.* Energoatomizdat, Moscow.

Meshalkin, G.S., N.P. Arkhipov, A.N. Arkhipov, and N.I. Burov. 1990. "Ratios, physical and chemical mobilities, bilogical accessibility, and migration characteristics for radioactive nuclides in the soil surrounding the Chornobyl NPP accident site." In *Chornobyl-90: papers from the first international conference on the biological and radioenvironmental aspects of the impact of the Chornobyl NPP accident* E.V. Senina, ed., pp. 253-266. Vol. 2 (*Radiation Genetics. Radiobiology. Agricultural Radiology*), Zelenyi Mys.

Ministry for Chornobyl Affairs. 1996. *Ukrainian national report on the 10th anniversary of the accident at ChNPP.* Ministry for Chornobyl Affairs, Kiev.

Moiseev, A.A., and L.V. Ramzaev. 1975. *Cesium-137 in the biosphere.* Atomizdat, Moscow.

Morozov, V.V., Ye.A. Konstantinov, Ye.M. Filippov, et al. 1990. *Analysis of FCM distribution, location condition and location data classification for the Shelter Facility.* VNIPIET Report No. 2669, Leningrad, (10 December).

National Council on Radiation Protection and Measurements (NCRP). 1985. *Induction of thyroid cancer by ionizing radiation.* NCRP, No.80, Bethesda, Maryland.

Nemchinov, Yu.I., N.G. Marenkov, A.M. Lisenyi, et al. 1997. "Evaluation of the structural condition of ChNPP Unit C and Phase II and recommendations for enhancement of seismic stability." In *Materials from the second international conference on the 10th anniversary of completion of work on Shelter Facility construction*, pp. 150-155. Ogni Slavuticha, Slavutich.

Nosko, B.S., B.S. Prister, M.V. Lobodi, et al. 1994. *Guide to agrochemical and archeological conditions in Ukrainian soils.* Urozhai, Kiev.

Novikova, S.K. 1989. "Vertical migration of radionuclides in soil with various types of contamination in the vicinity of ChNPP." In *Abstracts of papers presented at the all-union conference on principles and methods in landscape-based and geochemical studies of radionuclide migration*, pp. 80. 13-17 November, Suzdal', Moscow.

Nuclear Energy Agency (NEA). 1995. *Chernobyl ten years on radiological and health impact.* Nuclear Energy Agency, Paris, France.

Ogorodnikov, V.I., and V.V. Kanivets. 1993. "Bottom deposits in the Kiev Reservoir." *Trudy UkrNIGMI* (245):180-194.

Panfilov, A.P., L.F. Belovodskii, and V.N. Grimmanovskii. 1989. *Assuring radiation safety during construction of the Shelter Facility at ChNPP*. IAEA/SM-40, International Symposium on Remediation Work in the Event of a Nuclear Accident or Radiological Emergency, 6-10 November 1989, Vienna, Austria.

Pavlenko, T.O. 1996. "Scientific Basis for a radiation protection system to protect the Ukrainian public from radon-222." Dissertation (abstract), Kiev University, Kiev.

Pavlenko, T.A., I.P. Los, and N.V. Aksenov. 1996. "Indoor $^{222}$Rn levels and irradiation doses in the territory of the Ukraine." *Radiation Measurements, Great Britain* 26(4):585-591.

Pavlotskaya, F.I. 1974. *Migration of radioactive global fallout products in soil.* Atomizdat, Moscow.

Petryaev, E.P., T.G. Ivanova, T.K. Morozova, and G.A. Sokolik. 1989. "Prediction of vertical migration for radionuclides in the ChNPP release for soils from typical landscapes in Belarus." In *Abstracts from the all-union conference on principles and methods in landscape-based and geochemical studies of radionuclide migration,* pp. 86, 13-17 November 1989, Suzdal', Moscow.

Petukhova, N.N. 1986. "Lithological and geochemical characteristics od derno-podzolic soils in the Belorussian Poles'e." Dissertation (abstract), Minsk.

Poyarkov, V.A. 1997. "Multicomponent risk in nuclear and radiological accidents." In *IAEA conference on effects of low doses of ionizing radiation.* International Atmoic Energy Agency, Seville, Spain.

Pretsh, G. 1995. *Analysis of a roof collapse accident for the Shelter Facility at the Chornobyl Nuclear Power Plant.* Brief Description of Report GRS-A-2241, GRS, Germay.

Prister, B.S., V.A. Shevchenko, and V.A. Kal'chenko. 1982. "Genetic effects of radionuclides on agricultural vegetation." *Uspekhi Sovremennoi Genetiki,* pp. 27-60.

Prister, B.S., N.A. Loshchilov, O.F. Nemets, and V.A. Poyarkov. 1991. *Fundamentals of agricultural radiology.* 2nd ed., Urozhai, Kiev.

Prister, B.S., G.P. Perepelyatnikov, and L.V. Perepelyatnikova. 1993. "Countermeasures used in the Ukraine to produce forage and animal food products with radionuclide levels below intervention limits after the Chernobyl accident." *Sci. of the Total Environ.* 137:183-198.

Prister, B.S., Yu. A. Ivanov, L.V. Perepelyatnikova, and V.A. Pronevich 1996a. "Issues related to the implementation of countermeasures in Ukrainian agriculture following the Chornobyl NPP accident." *Vestnik Agrarnoi Nauki* (1):74-81.

Prister, B.S., Yu.O. Ivanov, L.V. Perepelyatnikova, M.N. Kudrik, A.O. Mozhar, and V.P. Skorobogat'ko. 1996b. "Agricultural aspects of the ChNPP accident cleanup."

In *The Chornobyl disaster*, ed. V.G. Bar'yakhtar, pp. 362-371. Naukova Dumka, Kiev.

Remis, V.P. 1996. "The application of cesium selective sorbents in the remediation and restoration of radioactive contaminated sites." In *Radioecology and the restoration of radioactive-contaminated sites*, Vol. 13, eds. Felix F. Luykx and Martin J. Frissel, NATO ASI Series, 2. Environment.

Remnick, D. 1993. *Lenin's tomb*. Random House, New York.

Repin, V.S. 1996. *Importance of sources for radiation hygiene and radiation doses received by the public following the accident at ChNPP*. PhD dissertation, Kiev.

Repin, V.S., O.A. Bondarenko, S.Yu. Nechaev, A.I. Bykorez, and L.I. Kononenko. 1993. "Dosimetry issues and risk assessment for inhalation of fuel particles. Current issues involving retrospective, current, and predicted radiation dosimetry related to the Chornobyl accident." In *Materials from the Ukrainian scientific conference*, pp. 23-34. 27-29 October 1992, Kiev.

Salbu B., T. Kerkling, D.H. Oughton, et al. 1994. "Hot particles in accidental releases from Chernobyl and Windscale nuclear installations." *Analyst* 119:125.

Salbu B., D.H. Oughton, V.A. Kashparov, S. Firsakova, and A.V. Konoplev. 1996. "Delayed mobilization of $^{90}$Sr associated with fuel particles." In *Book of extended synopses international conference "One decade after Chernobyl: Summing up the consequences of the accident."* CN-63/192, pp. 294-295. 8-12 April 1996, Austria Center, Vienna, Austria.

Sanzharova N.I., S.V. Fesenko, R.M. Alexakhin, V.K. Anisimov, V.K. Kuznetsov, and L.G. Chernyayeva. 1994. "Changes in the forms of $^{137}$Cs and its availability for plants as dependent on properties of fallout after the Chernobyl Nuclear Power Plant accident." *Sci. Total Environ.* 154:9-22.

Sharovarov, G.A., V.V. Kotovich, V.G. Molodykh, et al. 1995. *Radioactive contamination of Belarus in the event of a structural collapse in ChNPP Unit 4*. Vestnik, Akad. Nauk Belarus. Ser. Fiz.-Tekh. Nauk (1).

Shestopalov, V.M., ed. 1996. *Atlas of Chornobyl Exclusion Zone*. Kartografiya, Kiev.

Skorobogatko, E.P., and S.I. Rybalko 1992. "Radiochemical laws governing migration mobility of radionuclides released in the Chornobyl accident." In: *Chornobyl-92, papers from the third all-union scientific and technical conference on Chornobyl NPP accident cleanup*, Vol. 1, Part 3 (*Migration of Radionuclides in Natural Media*), pp. 465-471. Zelenyi Mys.

Slade, D. H., ed. 1968. *Meteorology and Atomic Energy—1968*, U.S. Atomic Energy Commission, Washington, D.C.

Sobolev, B., I. Likhtarev, I., Kairo, et al. 1996. "Radiation risk assessment of the thyroid cancer in Ukrainian children exposed due to Chernobyl, The radiological consequences of the Chernobyl accident, Minsk, Belarus," In *Proc. of the Confer-*

*ence*, pp. 741-748. Luxembourg, Office for Official Publications of the European Communities, March 18-22.

Tataurov, A.L. 1987. "Summary of CACH2 code." *Vopr. Atom. Nauk. Tekh. Ser. Fiz. Tekh. Atom. Reaktorov* (1):3-6.

Ukrainian Academy of Sciences. 1992. *Description of the Shelter Facility and requirements for its conversion*. Ukrainian Ministry for Remediation of the Chornobyl Accident, Ukrainian Academy of Sciences, Naukova Dumka, Kiev.

Ukrainian-Canadian Expedition to Reservoirs Along Dnepr River. 1995. *Comprehensive evaluation of heavy-metal, radionuclide, and pesticide pollution in water and bottom deposits associated with reservoirs along the Dnepr River*. UkrNDGMI, Kiev.

Ukrainian Unit. 1995. *Results obtained by the Ukrainian-canadian expert commission to study the current environmental status of the Dnepr River*. Ukrainian Unit, Khar'kov-Kiev.

UNIISKhR. 1998. *Improved determination of the strontium plume configuration in the area adjacent to ChNPP, and detailed study of the radiation environment in areas to be rehabilitated*. UNIISKhR, Kiev.

United Nations. 1982. *Ionizing radiation: sources and biological effects*. United Nations Special Commision on the Effects of Atomic Radiation, United Nations, New York.

United Nations Special Commission on the Effects of Atomic Radiation (UNSCEAR). 1993. *Sources and effects of ionizing radiation: UNSCEAR 1993 report to the general assembly*. United Nations Special Commission on the Effects of Atomic Radiation, United Nations, New York.

Usaty, A.F., et al. 1991. *Large scale gamma field distributions in the central hall of the Shelter Facility*. Kurchatov Institute Report No. 11.07/159, Kurchatov Institute, Moscow.

Vakulovskii, S.M., O.V. Voitsekhovich, et al. 1989. "Radioactive contamination of bodies of water in regions affected by the accident at ChNPP." In *Proceedings of the 1989 IAEA symposium*, pp. 231-246. International Atomic Energy Agency, Vienna, Austria.

Vakulovsky, S.M., A.I. Nikitin, and O. Voitsekhovitch. 1994. "Cesium-137 and strontium-90 contamination of water bodies in areas affected by releases from Chernobyl Nuclear Power Plant accident, an overview." *J. Environ.Radioactiv.* 23:103-122.

Vil'gusevich, I.P. 1955. "Leaching of potassium from derno-podzolic and sandy soil in the Belorussian SSR." *Trudy Inst. Sotsialist. Sel'skokhozyaistva Akad. Nauk BSSR* (3):107-115.

Voitsekhovich, O.V. 1992. *Improvement of radiation monitoring system for reservoirs on the Dnepr River and for bodies of water in the region affected by the Chornobyl accident. Analysis and study of abiotic factors and processes affecting migration of radioactive material into reservoirs on the Dnepr River to determine the self-purification capacity of reservoirs and predict specific effects resulting from their accidental contamination.* Vol. 1, Ministry of Chornobyl Affairs, Kiev.

Voitsekhovich, O.V. 1995. *Scientific validation and methodological support for integrated monitoring of radioactive contaminant outflow from the Exclusion Zone, and the hydrological transport and geochemical transformation of radioactive contaminants.* Vol. 2, Appendix 6, pp. 54-73.

Voitsekhovitch O., V. Kanivets, et al. 1996a. "Experimental studies of the radionuclides flux from rivers and radionuclide-sediment interaction in the Black Sea." In *Final report on International Atomic Energy Agency Research Contract 7330/R2/RB*, Kiev.

Voitsekhovitch O., V. Kanivets, I. Biliy, et al. 1996b. "Peculiarities of the particulate [137]Cs transport and sedimentation in Kiev Reservoir." *In Proceedings of the first international conference*, pp.179-183. 18-22 March 1996. Minsk, Belarus.

Yakushev, B.I., B.S. Martinovich, L.I. Rakhteenko, T.A. Budkevich, G.I. Kabashnikova, O.O. Ermakova, and M.M. Sak. 1996. "Circulation of radionuclides from the Chornobyl disaster in natural vegetation complexes under Belarusian conditions." In *International conference-one decade after Chernobyl: Summing up the consequences of the accident*, pp. 158-162. 8-12 April 1996. Vienna, Austria.

доблавение

# APPENDIX A
*Review of Radiation, Radioactivity,*
*and Radiation Protection*

## CONCEPTS AND BASIC DEFINITIONS

An individual element (e.g., hydrogen, carbon) has **isotopes** which can be described by the name of the element and the mass number (e.g., sodium-24 or $^{24}Na$). Isotopes of the same element behave the same chemically, but differ in the number of neutrons in their nuclei. When reference is made to the isotopes of more than one element, the correct term is **nuclide**. Some nuclides are unstable and undergo nuclear transformation to achieve a more stable configuration. These are referred to as **radionuclides**. During this transformation charged particles such as alpha and beta particles may be emitted. Sometimes these charged particle emissions are accompanied by the release of excess energy from the nucleus in the form of gamma radiation. Such transformations are callled **radioactive decay**.

**Activity** is the rate at which spontaneous nuclear transformations occur in a given amount of radioactive material. Activity is expressed in a unit called the **becquerel,** symbol **Bq**, where 1 Bq equals one nuclear transformation per second. The unit is named after the French physicist Henri Becquerel, who discovered radioactivity in 1896. Because this unit is so small, multiples (kilo- (k), x 1000; mega- (M), x $10^6$) are often used. The historical unit of radioactive transformation is the **curie**, symbol **Ci**, and was originally defined as the rate of nuclear transformations equal to that contained in one gram of $^{226}Ra$. This unit was named for the Polish scientist Marie Curie, who first isolated radium. The unit was later redefined as 3.7 x $10^{10}$ transformations per second. Because of the wide range of activity encountered in radioactive sources both submultiples (milli- (m), x $10^{-3}$; micro- ($\mu$), x $10^{-6}$) and multiples are used.

245

The time taken for the activity of a given quantity of a radionuclide to fall to half of its original value is called the **half-life** and is abbreviated as $t_{1/2}$. This value is unique to each radionuclide and is a function of the relative instability of the nuclear structure. Because of the different half lives, chemical behavior, and radiation types emitted by various radionuclides, it is not appropriate to add, for example, x becquerels of radionuclide A with y becquerels of radionuclide B.

When ionizing radiation passes through matter, a series of electrical interactions with the orbital electrons of atoms result. The primary results of this physical interaction are **ionization**, the production of a charged nucleus and free electron, or **excitation**, the imparting of energy to the orbital electrons without producing ionization. When ionizing radiation passes through cells or tissues in biological systems it can interact in a number of ways. The most significant of these are the formation of chemical **free radicals** such as the formation of the hydroxyl radical [$OH^-$] and hydronium [$H_3O^+$] from two molecules of water [$H_2O$] or interaction with the DNA in the nucleus of the cell. Free radicals can significantly alter biochemical processes. Interaction with the DNA molecule can result in temporary damage (e.g., single strand breaks) or permanent damage (e.g., multiple strand breaks.) Single breaks in the long double helical strands of the DNA molecule occur by many natural processes and are almost always immediately repaired. Multiple strand breaks in the DNA structure can also be repaired; however, the repair is often incomplete or incorrect. In the vast majority of cases, this results in the death of the single affected cell. In a very small number of cases, the damaged DNA is able to replicate and the result is a mutated cell. Some mutated cells can divide in an uncontrolled manner. In this case, the mutation can result in cancer. Questions about the repair and mutation processes of DNA are at the forefront of current research in molecular biology.

Ionizing radiation cannot normally be detected by the human senses. It can only be detected by observing changes in physical properties of material such as the darkening of photographic film or the production of electrical pulses in radiation detectors. From a physical standpoint, it is useful to measure the amount of energy absorbed in a given amount of matter. This quantity is called **absorbed dose** and the unit is called the **gray (Gy)**, which is equal to 1 joule per kilogram (J/kg). Because of its relatively large size, particularly in relation to occupational and environmental radiation protection standards, submultiples are often used. The historical unit of absorbed dose is the **rad**[1],

---

[1]The *rad* is both the historic unit name and the unit symbol for absorbed dose. If it is used as the unit name, it should be pluralized (e.g., 2 rads); however, it is invariant if used as the unit symbol.

where 1 rad is equal to the absorption of 100 erg/g (0.01 Gy). Different radiation types (i.e., alpha, beta, and gamma) do not necessarily have the same biological consequences given the same absorbed dose. The relative magnitude of the biological effect is related to the ionization density or linear energy transfer (LET) of the exposing radiation. Because of its large mass, the alpha particle does not have sufficient energy to penetrate a sheet of paper or the dead layer of skin covering the body. If an alpha-emitting radionuclide is taken into the body and alpha particles are emitted internally, they can produce more harm to a single cell than other types of radiation because they deposit all of their energy within a relatively small volume. For example, an alpha particle may travel only 40 or 50 $\mu$m in tissue, while a beta particle may travel 1 cm or more and a gamma photon may travel 10 cm or more.

To account for the differences in the way different radiation types interact with tissue, the quantity **dose equivalent** was developed. This quantity is the product of the absorbed dose multiplied by a dimensionless **weighting factor** (formerly known as quality factor) to account for the type of radiation involved. Basic standards for occupational and environmental radiation protection are typically expressed in terms of dose equivalent. The unit of dose equivalent used throughout this book is the **sievert (Sv)**[2] and is also equal to 1 J/kg. Similar to the gray, submultiples of the sievert are often used because of its relatively large size. Unlike the gray, this quantity cannot be measured directly and can only be calculated. The historical unit of dose equivalent is the **rem**[3], equal to the dose in rads multiplied by an appropriate weighting or quality factor (0.01 Sv). In 1990, this quantity was redefined as **equivalent dose** and the concept of quality factor was superceded by the **radiation weighting factor**. At the same time a new quantity, **effective dose**, was established to further account for the different radiation sensitivity of various body tissues to cancer, genetic effects, and overall health detriment. Unfortunately, all three of these quantities are expressed in the unit of sievert. This has caused a great deal of confusion, especially in nonEnglish speaking situations where there is essentially no distinction between the written form of the three terms. As this book is going to press, the International Commission on Radiological Protection (ICRP) is in the process of reviewing its basic recommendations. Hopefully any new recommendations will remedy this situation. Fortunately the authors have presented their results in the original quantity, dose equivalent.

---

[2]While the gray and sievert are both named for people, the unit names are always lowercase unless used as the first word in a sentence.

[3]Originally derived from an acronym for "roentgen equivalent man," the rem is both the historic unit name and unit symbol and follows the same rules for usage as the rad.

## DOSE STANDARDS FOR RADIATION PROTECTION

As our understanding of the biological effects of ionizing radiation has evolved, so has the endpoint of radiation protection. From the 1940s to early 1960s, a major concern in radiation protection was the prevention of genetic effects. As studies of the major radiation dose populations—atomic bomb survivors, radium dial painters, and other selected high dose populations—progressed and matured, it became apparent that genetic effects were absent. The primary concern shifted to the prevention of life-shortening disease, particularly cancer.

Standards for radiation protection at the time of the Chornobyl accident were a combination of ICRP recommendations from 1959–1976. In the case of uniform exposure to the whole body, the occupational radiation protection standard was an annual dose equivalent limit of 50 mSv. For cases involving nonuniform irradiation, standards included weighting factors that take into account differences in the relative radiation sensitivity of various tissues. The permissible exposure limit for the general population was set at 5 mSv/yr. These limits include doses received from exposure to external sources as well as the tissue weighted doses from radioactive materials taken into the body through inhalation, ingestion, or absorption and, therefore, are sometimes referred to as **total effective dose equivalent (TEDE)**.

The dose from the intake of a given quantity of radioactive material depends upon the type of radiation emitted, the half-life of the radionuclide, and the chemical characteristics of the element or compound. For controlling doses to workers from intakes of radioactive materials, an **annual limit on intake (ALI)** expressed in becquerels has been established for each radionuclide. An intake of activity equal to the ALI for a radionuclide will result in a **committed effective dose equivalent (CEDE)** of 50 mSv.[4] The CEDE concept takes into account that once taken into the body, a radionuclide is transported to organs or tissues where it remains until it is eliminated by a combination of excretion and radioactive decay. For short-lived radionuclides, such as $^{131}I$, the entire dose is received in a few months following an intake. In the case of longer-lived radionuclides, specifically long-lived bone seekers such as $^{90}Sr$ or $^{239}Pu$, the dose to organs and tissues from the intake of a radionuclide may occur over a period of years. To simplify dose and compliance calculations, the entire dose to the worker is assumed to be received in the year in which the intake occurred. This simplifying assumption means that while

---

[4]Note: The ALI values derived on the basis of the 1976 ICRP recommendations are based on the 50 mSv standard. ALI values based on the 1991 recommendations are based on a 20 mSv standard. In Ukraine the 50 mSv limit was in effect from the time of the accident through 1999. The 20 mSv limit for radiation workers in Ukraine takes effect in 2000.

some of the dose from an intake of a radionuclide may occur in later years, this entire dose is "charged" against the worker's dose limit for the year in which the intake occurred.

While the ALI values for each radionuclide are unique, the intake expressed as $<$ fraction of the ALI for a particular radionuclide may be added in the case of multiple radionuclides to demonstrate compliance with the radiation protection standard:

$$\sum_i \frac{A_i}{ALI_i} < 1$$

where $A_i$ is the activity of the $i^{th}$ radionuclide and $ALI_i$ is the ALI for the $i^{th}$ radionuclide.

If the sum of the frac$<$ ons equals unity, then the committed effective dose equivalent is 50 mSv. If external and internal doses are considered, the following equation demonstrates compliance with the radiation protection standard:

$$\frac{H_E}{50mSv} + \sum_i \frac{A_i}{ALI_i} < 1$$

where $H_E$ is the external dose equivalent .

The notation used by the ICRP, the authors, and in this appendix varies somewhat; however, the basic concepts remain the same.

The authors also use an earlier and more complex standard for protection against the intake of radioactive materials—the **maximum permissible concentration (MPC).** If a worker inhales (or drinks) a concentration of radioactive material equal to the MPC for 2,000 hours per year for 50 years, the worker will receive a dose equal to the maximum permissible dose for a particular organ or the total body. These limits are based on the previous (1959) radiation protection recommendations of the ICRP.

To place these limits into perspective, the average dose to a member of the public in the United States from all sources of ionizing radiation is 3.6 mSv /yr (360 mrem/yr; approximately 1 mrem/d). This includes the dose from natural sources such as radon and cosmic radiation, which account for 82% of the total, as well as the dose from technogenic sources including medical doses, consumer products, weapons testing fallout, and nuclear power.

## RADIATION EXPOSURE AND HEALTH EFFECTS

Harmful effects from exposure to radiation can be broadly divided into two categories: early and late. Early effects typically result from high doses at high dose rates (i.e., acute doses) where the rate of injury to tissues or organs exceeds the rate of repair. These effects typically involve some threshold. For example, clinical symptoms are observed following an acute whole body dose of 1 Gy. Table A.1 summarizes the acute effects of radiation exposure to the whole body.

Partial body exposure is a more common scenario in radiation accidents, especially those involving the handling or contact with small radioactive sources. In such cases, the primary effect of partial body irradiation is to the skin and underlying tissues. Table A.2 summarizes the effects of acute radiation exposure to the skin.

**TABLE A.1.** Summary of Acute Effects of Whole Body Irradiation

| Dose Range, Gy | Effects | Prognosis |
|---|---|---|
| <1 | Subclinical | |
| 1–8 | Hematopoietic syndrome —acute damage to bone marrow resulting in severe depression of lymphocytes; greatly reduced ability to fight infection | < 4 Gy—good<br>4 Gy—50% lethality in 30 d<br>6 Gy—90% lethality |
| 8–9 | Respiratory syndrome—radiation pneumonitis | Lethal in 14 - 30 d |
| 8–30 | Gastrointestinal syndrome—destruction of the crypt cells lining the small intestine; electrolyte imbalance; intestinal bleeding; sepsis | Lethal in 2 wk |
| >30 | Cardiovascular/central nervous syndrome— disorientation; unconsciousness; cerebral edema; cardiovascular collapse | Lethal in 72 h |

Sources: UNSCEAR [1988] and Wald [1992]

**TABLE A.2.** Summary of Acute Effects of Skin Irradiation

| Dose Range (Gy) | Effect |
|---|---|
| <3 | Subclinical |
| 3 | Epilation—loss of hair |
| 6 | Erythema—reddening of the skin |
| 10–20 | Dry desquamation—blistering |
| 30–50 | Wet desquamation—ulceration |
| >50 | Necrosis—gangrene |

While the surviving victims of acute effects are also presumed to be at greater risk of late effects, such effects are typically associated with lower doses and lower dose rates. The primary late effect of chronic radiation exposure is cancer. Functional damage to organs and tissues can result from high doses at any rate. A well-established example of such an effect is the development of cataracts in workers who have received large doses to the eyes. Cataracts are sometimes observed among patients treated for cancers where treatment has involved irradiation of the head. Significant effects of radiation dose to the developing fetus include spontaneous abortion, and, in fetuses carried to term, severe mental retardation. Cleft palate and other developmental disabilities, and childhood leukemia have also been observed. Maximum sensitivity for teratogenic effects occurs 8–15 weeks post-conception, the period of organ differentiation and initial development.

Sensitivity to radiation exposure is a complex function of gender, age at exposure, radiation type, dose rate, and other factors such as general health and nutrition.

## REFERENCES

United Nations Scientific Committee on the Effects of Atomic Radiation (UNSCEAR). 1988. *Sources and effects of ionizing radiation.* UNSCEAR 1988 Report, United Nations, New York.

Wald, N. 1992. "Acute radiation injuries and their medical management." Chapter 12 in The biological basis of radiation protection practice in eds. K. Mossman and W. Mills, Williams and Wilkins, Baltimore.

# APPENDIX B

*Medical Consequences of the
Chornobyl Accident and Implications
for Human Health Risk Assessment*

The medical consequences of the Chornobyl Nuclear Power Plant accident have often been the subject of wild and absurd claims. A comprehensive treatment of the risks posed by the Chornobyl accident and its aftermath would be incomplete without some examination of these consequences and their implication for human health risk assessment. Some accounts place the number of accident-related fatalities at levels equivalent to nearly all of the deaths from all causes in the affected regions of Ukraine, Belarus, and Russia since 1986. While there are many excellent publications dealing exclusively with the human health dimensions of the Chornobyl accident, this appendix is intended to compliment the information in Chapter 6. In particular, four significant medical effects are reviewed:

- Acute radiation syndrome

- Thyroid cancer

- Leukemia

- Teratogenesis

There are relatively few sources of data that can be used to evaluate the human health risks posed by ionizing radiation. Before the Chornobyl accident, the largest population was the survivors of the atomic bombings of Hiroshima and Nagasaki. Much of the current philosophy of radiation protection and human health risk estimates are based on these individuals and other much smaller groups of significantly exposed individuals such as radium dial painters. For the most part, these individuals received relatively high doses at high dose rates. Our understanding of these acute effects is based on a relatively small population. Radiation accidents rarely involve uniform irradiation of the whole body.

This summary of acute radiation syndrome is based on the 1988 United Nations Scientific Committee on the Effects of Atomic Radiation Report (UNSCEAR 1988a; UNSCEAR 1988b) and summaries prepared by the International Atomic Energy Agency (IAEA) (1988). The discussion of thyroid cancers is based primarily on reports by Stsjazhko et al. (1995) and Likhtarev et al. (1994). The summary on leukemia is based on the 1986 statement of the IAEA as supported by subsequent publications including Ivanov et al. (1997) and Gluzman et al. (1999). The discussion on teratogenesis is based on a review by Castronovo (1999).

## ACUTE RADIATION SYNDROME

Acute effects are those that occur within hours to weeks following a high absorbed dose, typically greater than 1 Gy to the whole body. These effects exhibit a threshold below which they are not observed, above the threshold their severity increases with increasing dose. These are sometimes called deterministic effects, although some deterministic effects such as cataracts can result from large doses accumulated over a much longer time. Acute effects may be life threatening depending on the magnitude of the dose and other factors such as the general health of the exposed individual and the effects of combined injury (e.g., burns). Victims of acute radiation exposure are often susceptible to infection that may prove fatal because of severely depressed immune response.

In the Chornobyl accident acute effects were limited to those personnel who were on-site at the time of the accident or called to fight the fires that followed the explosion of the Unit 4 reactor. Two individuals were killed immediately (one of whom was never recovered) and another died of thermal burns in the day following the accident. When combined with the 28 fatalities that occurred in the months following the accident, there were a total of 31 early fatalities attributed either in whole or in part to acute radiation syndrome.

Each of the affected individuals was hospitalized, with the more severely exposed cases treated in Moscow. Table B.1 shows the distribution of doses among these individuals and the outcomes. The severity and speed of onset of symptoms was dependent on the magnitude of the dose. Initial symptoms included diarrhea, vomiting, and erythema (reddening of the skin). Approximately 200 individuals were hospitalized either in regional hospitals or specialized centers in Moscow and Kiev. No members of the general public received doses sufficient to result in acute radiation syndrome (ARS). Careful monitoring of the differential white cell count (lymphocytes) was

**TABLE B.1.** Acute Radiation Syndrome Cases from the Chornobyl Nuclear Power Plant Accident

| Acute Radiation Syndrome Degree | Estimated Dose (Gy) | Treatment Center | | Deaths | |
|---|---|---|---|---|---|
| | | Moscow | Kiev | Total | % |
| IV | 6–16 | 20 | 2 | 20 | 91 |
| III | 4-6 | 21 | 22 | 7 | 16 |
| II | 2-4 | 43 | 10 | 1 | 2 |
| I | 1-2 | 31 | 74 | 0 | 0 |

Reference: IAEA 1988.

used to classify the severity of exposure (degrees I through IV). Because of technical limitations of the dosimetry system used at Chornobyl at the time of the accident, dose estimates are based on a combination of biological dosimetry and clinical symptoms. Lymphocyte counts were the primary means of assessing doses; however, chromosome aberration analysis was used in higher dose cases, particularly degrees III and IV.

Degree IV ARS is characterized by a rapid onset of symptoms, less than 1 hour following exposure. The lymphocyte count in these individuals was observed to fall below 100 per $\mu$L in the first week (normal range 1,000 to 3,000 per $\mu$L). From a normal platelet count of 150,000 to 400,000 per $\mu$L, levels as low as 500 $\mu$L were observed 8 to 9 days after exposure. Individuals in this category had both severely compromised immune response as well as greatly reduced blood-clotting capability. They experienced severe diarrhea and electrolyte imbalance. Mucous membranes were particularly sensitive, making breathing and swallowing difficult. Skin burns from beta radiation complicated management of these injuries. Thermal burns further confounded two cases. Only 2 of 22 patients in this category survived despite heroic medical measures.

Degree III ARS is also characterized by an early onset of symptoms (i.e., nausea and vomiting) within an hour of exposure. The minimum leukocyte count was observed to be as low as 200 per $\mu$L. Patients in this category also presented with toxemia and fever. These individuals also experienced skin burns from beta radiation and significant inflammation of mucous membranes. Following heroic treatment similar to that given to the degree IV ARS patients including reverse isolation and antibiotic therapy, this group fared significantly better than those experiencing degree IV injury. The survival rate 84%.

Patients with degree III and IV injury were also treated with intravenous feeding to prevent infection through the gastrointestinal tract and to manage electrolyte balance. Patients with doses above 2 Gy were also treated with antibiotics and antifungal agents. Platelet transfusions were administered in cases where the patient's count dropped below 1,000 per $\mu$L.

Degree II patients typically experience symptoms 1 to 2 hours following exposure. The lymphocyte count in this group typically decreased below 500 per $\mu$L, but remained above 200 per $\mu$L. Patients in this group also presented with toxemia and fever; however, the beta radiation burns were not as significant. These individuals also experienced a significant decrease in the number of blood platelets; however, the minimum level occurred much later —at approximately 21 days. The survival rate for this group was more than 95%.

Degree I patients experienced much less severe symptoms with an onset several hours following exposure. Two patients in this category also suffered extensive burns. As expected, the survival rate for this group was 100%.

Thirteen patients with doses judged to be above 4 Gy received bone marrow transplants. Eleven of these patients subsequently died. The results of this experience indicate that even at relatively high doses, bone marrow may not be completely destroyed. In such cases a "host versus graft" reaction likely occurred. The uncertainty in the dosimetry for these individuals – the basis for selecting bone marrow transplant candidates – was another significant contributing factor in the limited effectiveness of this treatment. A major lesson of this experience was that bone marrow damage is probably best managed by the prompt administration of hematopoietic growth factors. The application of recent advances in cytokine development for cancer therapy combined with stem cell transplants using fetal cord blood now appears to be the treatment of choice for such cases.

## THYROID CANCER

The primary medical consequence of the Chornobyl accident was a dramatic increase in child thyroid cancer resulting from the consumption of milk contaminated with $^{131}$I during the months immediately following the accident. Stsjazhko et al. (1995) reported dose estimates for approximately 100,000 children in Ukraine and Belarus based on in vivo measurements. These results are shown in Table B.2. Of the children measured, 57% had doses less than 300 mGy. Approximately 7% had doses greater than 2 Gy and 0.6% (641 children) had doses greater than 10 Gy. Studies to provide a more comprehensive assessment of the child thyroid doses are ongoing (Likhtarev et al. 1994; Heidenreich et al. 1999; Jacob et al. 1999). Most of the dose reconstruction is based on environmental measurements of $^{137}$Cs.

**TABLE B.2.** Preliminary Estimates of Thyroid Doses Among Children in Ukraine and Belarus Based on In Vivo Measurements

| Estimated Thyroid Dose (mGy) | Belarus | | Ukrane | |
|---|---|---|---|---|
| | Number | Percentage | Number | Percentage |
| 0-300 | 13,556 | 49.8 | 45,938 | 60.3 |
| 300-1,000 | 8,631 | 31.7 | 19,293 | 25.3 |
| 1,000-2,000 | 2,808 | 10.3 | 5,684 | 7.5 |
| 2,000-5,000 | 1,743 | 6.4 | 3,698 | 4.9 |
| 5,000-10,000 | 370 | 1.4 | 1,012 | 1.3 |
| >10,000 | 111 | 0.4 | 530 | 0.7 |
| Total | 27,217 | 100 | 76,155 | 100 |

Reference: Stsjazhko et al. (1995)

The difference in the time course of the release, chemical behavior, and deposition between cesium and iodine introduces significant uncertainty into such estimates.

The normal incidence of child thyroid cancer is low, typically less than 0.5 cases per million. Table B.3 shows the change in reported cancer cases before and following the Chornobyl accident. The post-accident incidence per million rose by as much as a factor of 200 from pre-1986 levels with an apparent latent period as short as 4-5 years. Tronko et al. (1999) report 577 cases in Ukraine between 1986 and 1997. Buglova et al. (1996) report that the incidence in boys is 50% higher than in girls. Goldman (1997) predicts that the ultimate number of cases will reach between 3,000 and 6,000.

## LEUKEMIA

An increase in the incidence of leukemia is a major concern following significant radiation exposure. According to analyses presented at the 1996 IAEA conference on the consequences of the Chornobyl accident (IAEA 1996), the total expected excess fatalities because of leukemia will be on the order of 470 among the 7.1 million inhabitants of contaminated areas and control zones. Superimposed on the expected spontaneous incidence of

**TABLE B.3.** Number and Incidence of Child Thyroid Cancer (Children Under 15 at Time of Diagnosis) in Belarus, Ukraine, and Russia Before and After the Chornobyl Accident

| | Number | | | Rate (per million) | | |
|---|---|---|---|---|---|---|
| Area | 1981–1985 | 1986–1990 | 1991–1994 | 1981–1985 | 1986–1990 | 1991–1994 |
| **Belarus** | 3 | 47 | 286 | 0.3 | 4.0 | 30.6 |
| Gomel Region | 1 | 21 | 143 | 0.5 | 10.5 | 96.4 |
| **Ukraine** | 25 | 60 | 149 | 0.5 | 1.1 | 3.4 |
| Northern 5 regions | 1 | 21 | 97 | 0.1 | 2.0 | 11.5 |
| **Russia** | NA | NA | NA | NA | NA | NA |
| Bruansk, Kaluga regions | 0 | 3 | 20 | 0 | 1.2 | 10.0 |

NA = data not available
Reference: Stsjazhko et al. (1995)

25,000 cases, these excess cases would be impossible to discern. The expected number of cases among an assumed 200,000 liquidators[1] would be on the order of 200 cases compared to a spontaneous incidence of 800 cases. Using current models, 150 of the 200 expected excess leukemia cases would have been expected in the first 10 years after the accident, following a latent period of 3 to 5 years. The spontaneous incidence in the liquidator population for the same period would be approximately 40 cases. To date, no consistent attributable increase in the rate of leukemia has been detected.

Ivanov et al. (1997) report no statistically significant difference in cancer morbidity or mortality between inhabitants of contaminated and uncontaminated areas of the Kaluga Oblast (region) in Russia. Gluzman et al. (1999) report that while childhood leukemia represents one of the most likely health effects in a significantly exposed population, no significant increase has been observed.

---

[1]The term applied to the workers who participated in early remediation activities at Chornobyl in 1986–1987.

Any meaningful conclusions on the link between the Chornobyl accident and leukemia are confounded by the poor quality of data on incidence rates in affected areas of Ukraine, Belarus, and Russia before 1986, selection bias for participation in health effect studies, and the general decline in nutrition and environmental quality throughout the countries of the former Soviet Union.

## TERATOGENESIS

Ionizing radiation is a well-established teratogen. Teratogenic effects are deterministic with a threshold of approximately 0.1 Gy (ICRP 1993; UNSCEAR 1986). Two specific effects are of concern: severe mental retardation and cancer that may be expressed in childhood or adult life.

The potential for organ malformation is highest in exposures occurring between the third through eighth weeks postconception. The natural incidence of such malformation is 0.06 or approximately 1 in 17 live births.

Irradiation during the eighth through twenty-fifth weeks postconception carries the potential for severe mental retardation. The effect is most significant in the eighth through fifteenth weeks with an observed shift of 30 intelligence quotient (IQ) points per Gy. Doses in the sixteenth through twenty-fifth weeks would show similar, but less severe effects per unit dose above the threshold.

The risk of childhood cancers following irradiation in utero is believed to be approximately the same as for infants and children irradiated in the first decade of life (ICRP 1993), with a 2–3 times higher risk than the general population.

Castronovo (1999) provides a thorough review of embryo/fetal studies conducted throughout Europe and Asia Minor. Many of the studies that do indicate some increase in fetal abnormality are flawed by poor data quality, bad experimental design, and selection bias. When such studies are discounted, there is no evidence that the radiation exposures of pregnant women arising from the Chornobyl accident produced any harmful effects.

There are, of course, many anecdotal accounts of birth defects and childhood cancers in Ukraine, Belarus, Russia, and other countries contaminated by the Chornobyl accident (Kotz 1995). While the reality of these cases cannot be discounted, the spatial and temporal association of a particular disease or congenital abnormality does not equal causation. As noted previously here and elsewhere, the effects of decreased nutrition, stress, environmental pollution from a variety of nonradiological sources, and a general lack of preventive (and prenatal) medical care must be taken into consideration when attempting to analyze the cause of such cases. This is not an argument for any theories of radiation hormesis, but simply an acknowledgment that the data are inconclusive and that ionizing radiation cannot be considered in isolation from other human health risk factors in such a complex and dynamic system.

It is noteworthy that major international accounts of the health effects of the Chornobyl accident are silent on one of its greatest socio-medical impacts—the dramatic increase in abortions in the Soviet Union and Central and Eastern Europe following the accident. The accounts of both the IAEA (1996) and Nuclear Energy Agency (1996) are utterly silent on this matter. While quantitative data on this was difficult to obtain, studies confirm that thousands of women throughout the Soviet Union, Central and Eastern Europe underwent abortions in the months and early years following the Chornobyl accident out of a fear of health effects in their children (Dewals and Dolk 1990; Czeizel 1991).

## CONCLUSION

It will be several decades before the human health consequences of the Chornobyl accident are fully understood. Even now, only 13 years after the accident, many observations and preliminary findings indicate that previously held models and assumptions about the effects of ionizing radiation might need to be revised. The experience gained in the treatment of the acute radiation syndrome cases greatly increased our understanding and improved techniques for managing major radiation injuries. The dramatic increase in child thyroid cancer reinforces the need for prompt intervention following a major radiological accident to control population doses. The apparent absence of leukemia cases among the 200,000 liquidators and the larger population in the affected areas of Ukraine, Belarus, and Russia raises some important questions about existing risk models and may highlight the importance of dose rate and genetic repair mechanisms as major factors in carcinogenesis. Ultimately, this may also result in revised risk estimates for some radiogenic diseases.

As our understanding of molecular biology increases, the knowledge gained through a careful study of the effects of the Chornobyl accident may also provide insights into human health risk factors beyond ionizing radiation. This great tragedy presents a unique opportunity to improve our understanding of both human health and ecological risk.

## REFERENCES

Buglova, E.E., J.E. Kenigsberg, and N.V. Sergeeva. 1996. "Cancer risk estimating in Belarussian children due to thyroid irradiation as a consequence of the Chornobyl nuclear accident." *Health Physics* 71:45-49.

Castronovo, F.P. 1999. "Teratogen update: radiation and Chernobyl." *Teratology* 60:100-106.

Czeizel, A.E. 1991. "Incident of legal abortions and congenital abnormalities in Hungary." *Biomedical Pharmacotherapy* 45:249-254.

Dewals, P., and W. Dolk. 1990. "Effects of the Chernobyl radiological contamination on human reproduction in Western Europe." *Progress in Clinical Biological Research* 340:339-346.

Gluzman, D.F., I.V. Abramenko, L.M. Sklyarenko, V.A. Nadgotnaya, M.P. Zavelevich, N.I. Bilous, and L.Y. Poludnenko. 1999. "Acute leukemias in children from the city of Kiev and Kiev region after the Chernobyl NPP catastrophe." *Pediatric Hematology and Oncology* 16:355-360.

Goldman, M. 1997. "The Russian radiation legacy." *Environmental Health Perspectives* 105:S6 1385-1391.

Heidenreich, W.F., J. Kenigsberg, P. Jacob, E. Buglova, G. Goulko, H.G. Paretzke, E.P. Demidchik, and A. Golovneva. 1999. "Time trends of thyroid cancer incidence in Belarus after the Chernobyl accident." *Radiation Research* 151:617-625.

International Atomic Energy Agency (IAEA). 1988. *Medical aspects of the Chernobyl accident.* IAEA TECDOC-516, International Atomic Energy Agency, Vienna, Austria.

International Atomic Energy Agency (IAEA). 1996. *International conference – One decade after Chernobyl: Summing up the consequences of the accident.* International Atomic Energy Agency, Vienna, Austria.

International Commission on Radiological Protection (ICRP). 1993. "Summary of the current ICRP principles for protection of the patient in nuclear medicine. A report by Committee 3 of the International Commission on Radiological Protection. International Commission on Radiological Protection." *Annals of the ICRP* 24(4):ix-x.

Ivanov, V.K., A.F. Tsyb, E.V. Nilova, V.F. Efendiev, A.I. Gorsky, V.A. Pitkevich, N.A. Leshakov and V.I. Shiryaev. 1997. "Cancer risks in the Kaluga oblast of the Russian Federation 10 years after the Chernobyl accident." *Radiation and Environmental Biophysics* 36(3):161-167.

Jacob, P., Y. Kenigsberg, I. Zvonova, G. Goulko, E. Buglova, W.F. Heidenreich, A. Golovneva, A.A. Bratilova, V. Drozdovitch, J. Kruk, G.T. Pochtennaja, M. Balanov, E.P. Demidchik, and H.G. Paretzke. 1999. "Childhood exposure due to the Chernobyl accident and thyroid cancer risk in contaminated areas of Belarus and Russia." *British Journal of Cancer* 80(0):1461-1469.

Kotz, D. 1995. "Investigating Chernobyl-induced thyroid cancer: politics versus science." *Journal of Nuclear Medicine* 36:15-16, 24, 29.

Likhtarev, I.A. V.V. Chumak, and V.S. Repin. 1994. "Retrospective reconstruction of individual and collective external gamma doses of population evacuated after the Chernobyl accident." *Health Physics* 66:643-652.

Nuclear Energy Agency. 1996. *The Chernobyl accident: 10 years on—radiological and health consequences*. Organization for Economic Cooperation and Development, Paris.

Stsjazhko, V.A., A.B. Tsyb, N.D. Tronko, G. Souchkevitch, and K.F. Baverstock. 1995. "Child thyroid cancer since accident at Chernobyl." *British Medical Journal* 310:801.

Tronko, M.D., T.I. Bogdanova, I.V. Komissarenko, O.V. Epstein, V.Oliynyk, I.A. Likhtarev, I. Kairo, S.B. Peters, and V.A. LiVolsi. 1999. "Thyroid carcinoma in children and adolescents in Ukraine after the Chernobyl nuclear accident." *Cancer* 86:149-156.

United Nations Special Commission on the Effects of Atomic Radiation (UNSCEAR). 1986. *UNSCEAR Report for 1986. United Nations Scientific Committee on the Effects of Atomic Radiation. Annex C—Biological effects of pre-natal irradiation*. United Nations Special Commission on the Effects of Atomic Radiation, United Nations, New York, pp. 263-366.

United Nations Special Commission on the Effects of Atomic Radiation (UNSCEAR). 1988a. *UNSCEAR Report for 1988. United Nations Scientific Committee on the Effects of Atomic Radiation. Annex D—Exposures from the Chernobyl accident*. New United Nations Special Commission on the Effects of Atomic Radiation, United Nations, New York, pp. 309-374.

United Nations Special Commission on the Effects of Atomic Radiation (UNSCEAR). 1988b. *UNSCEAR Report for 1988. United Nations Scientific Committee on the Effects of Atomic Radiation. Annex G—Early effects in man of high doses of radiation*. United Nations Special Commission on the Effects of Atomic Radiation, United Nations, New York, pp. 545-647.

доблавение

# APPENDIX C
*Acronyms and Abbreviations*

| | |
|---|---|
| AMAD | activity mean aerodynamic diameter |
| ICRP | International Commission on Radiological Protection |
| MinSredMash | Ministry of Mid-Level Industrial Engineering |
| MPC | maximum permissible concentrations |
| MPD | maximum permissible dose |
| NPP | nuclear power plant |
| RBMK | high powered channelized reactor |
| RCP | reactor coolant pump |
| RCPS | reactor control and protection system |
| TNT | trinitrotoluene |
| UkrNIGMI | Ukrainian Scientific Research Institute for Hydrology and Meteorology |
| USSR | Union of Soviet Socialist Republics |
| VNIPIET | All-Union Scientific Research, Design, and Experimental Design Institute for Power Technology |

# INDEX